A SURVEY OF HIGH-LEVEL SYNTHESIS SYSTEMS

THE KLUWER INTERNATIONAL SERIES IN ENGINEERING AND COMPUTER SCIENCE

VLSI, COMPUTER ARCHITECTURE AND DIGITAL SIGNAL PROCESSING

Consulting Editor
Jonathan Allen

Latest Titles

Analog VLSI Implementation of Neural Systems, C. Mead, M. Ismail (Editors)
ISBN: 0-7923-9040-7
The MIPS-X RISC Microprocessor, P. Chow, ISBN: 0-7923-9045-8
Nonlinear Digital Filters: Principles and Applications, I. Pitas, A. N. Venetsanopoulos, ISBN: 0-7923-9049-0
Algorithmic and Register-Transfer Level Synthesis: The System Architect's Workbench, D.E. Thomas, E.D. Lagnese, R.A. Walker, J.A. Nestor, J.V. Ragan, R.L.Blackburn, ISBN: 0-7923-9053-9
VLSI Design for Manufacturing: Yield Enhancement, S.W..Director, W. Maly, A.J. Strojwas, ISBN: 0-7923-9053-7
Testing and Reliable Design of CMOS Circuits, N.K. Jha, S. Kundu, ISBN: 0-7923-9056-3
Hierarchical Modeling for VLSI Circuit Testing, D. Bhattacharya, J.P. Hayes, ISBN: 0-7923-9058-X
Steady-State Methods for Simulating Analog and Microwave Circuits, K. Kundert, A. Sangiovanni-Vincentelli, J. White, ISBN: 0-7923-9069-5
Introduction to Analog VLSI Design Automation, M. Ismail, J. Franca, ISBN: 0-7923-9102-0
Gallium Arsentide Digital Circuits, O. Wing, ISBN: 0-7923-9081-4
Principles of VLSI System Planning, A.M. Dewey ISBN: 0-7923-9102-0
Mixed-Mode Simulation, R. Saleh, A.R. Newton, ISBN: 0-7923-9107-1
Automatic Programming Applied to VLSI CAD Software: A Case Study, D. Setliff, R.A. Rutenbar, ISBN: 0-7923-9112-8
Models for Large Integrated Circuits, P. Dewilde, Z.Q. Ning ISBN: 0-7923-9115-2
Hardware Design and Simulation in VAL/VHDL, L.M. Augustin, D.C..Luckham, B.A.Gennart, Y.Huh, A.G.Stanculescu ISBN: 0-7923-9087-3
Subband Image Coding, J. Woods, editor, ISBN: 0-7923-9093-8
Low-Noise Wide-Band Amplifiers in Bipolar and CMOS Technologies,Z.Y.Chang, W.M.C.Sansen, ISBN: 0-7923-9096-2
Iterative Identification and Restoration of Images, R. L.Lagendijk, J. Biemond ISBN: 0-7923-9097-0
VLSI Design of Neural Networks, U. Ramacher, U. Ruckert ISBN: 0-7923-9127-6
Synchronization Design for Digital Systems, T. H. Meng ISBN: 0-7923-9128-4
Hardware Annealing in Analog VLSI Neurocomputing, B. W. Lee, B. J. Sheu ISBN: 0-7923-9132-2
Neural Networks and Speech Processing, D. P. Morgan, C.L. Scofield ISBN: 0-7923-9144-6
Silicon-on-Insulator Technology: Materials to VLSI, J.P. Colinge ISBN: 0-7923-9150-0
Microwave Semiconductor Devices, S. Yngvesson ISBN: 0-7923-9156-X

A SURVEY OF HIGH-LEVEL SYNTHESIS SYSTEMS

edited by

Robert A. Walker
Rensselaer Polytechnic Institute

and

Raul Camposano
IBM

Kluwer Academic Publishers
Boston/Dordrecht/London

Distributors for North America:
Kluwer Academic Publishers
101 Philip Drive
Assinippi Park
Norwell, Massachusetts 02061 USA

Distributors for all other countries:
Kluwer Academic Publishers Group
Distribution Centre
Post Office Box 322
3300 AH Dordrecht, THE NETHERLANDS

Library of Congress Cataloging-in-Publication Data

A Survey of high-level synthesis systems / edited by Robert A. Walker and Raul Camposano.
p. cm. — (The Kluwer international series in engineering and computer science. VLSI, computer architecture, and digital signal processing.)
Includes index.
ISBN 0-7923-9158-6 (alk. paper)
1. Integrated circuits—Very large scale integration—Design and construction—Data processing. 2. Silicon compilers. I. Walker, Robert A., 1959- . II. Camposano, Raul. III. Series.
TK7874.S857 1991
621.39'5—dc20 91-14270
CIP

Printed on acid-free paper.

Printed in the United States of America

to our parents

and to Ellen, from Bob

Table of Contents

Preface

After long years of work that have seen little industrial application, high-level synthesis is finally on the verge of becoming a practical tool. The state of high-level synthesis today is similar to the state of logic synthesis ten years ago. At present, logic-synthesis tools are widely used in digital system design. In the future, high-level synthesis will play a key role in mastering design complexity and in truly exploiting the potential of ASICs and PLDs, which demand extremely short design cycles.

Work on high-level synthesis began over twenty years ago. Since then, substantial progress has been made in understanding the basic problems involved, although no single universally-accepted theoretical framework has yet emerged. There is a growing number of publications devoted to high-level synthesis, specialized workshops are held regularly, and tutorials on the topic are commonly held at major conferences.

This book gives an extensive survey of the research and development in high-level synthesis. In Part I, a short tutorial explains the basic concepts used in high-level synthesis, and follows an example design throughout the synthesis process. In Part II, current high-level synthesis systems are surveyed.

The task of selecting the systems to include in the survey was by no means an easy one. There is a fairly strong consensus today on exactly what the term "high-level synthesis" means, and only the systems that fit this definition were included. However, any restriction by topic always raises border cases. For this survey, systems that convert Register-Transfer level descriptions into logic level descriptions (often called "silicon compilers") were not included; neither were "system level" design tools.

The second criterion used for inclusion in the survey was a reasonable publication record. For the most part, we chose the arbitrary limit of two publications in conferences, journals or magazines in the field. Determining exactly which conferences, journals and magazines are considered to be "in the field" (automatic synthesis of digital circuits) is of course subjective; the ones we included will certainly raise no discussion, while the ones we did not pay enough attention to, may.

The last criterion applied to select systems for inclusion was timeliness. We reduced this criterion to a single number: with a few exceptions of historical value, no systems without publications after 1985 were included. Since 1985, sessions with titles such as "Synthesis of Logic Structures from Behavioral Descriptions", "Control/Data Path Synthesis", "High-Level Synthesis", etc., have become numerous in major conferences, leading to our choice of that cut-off date.

The book is intended to provide a convenient reference on high-level synthesis. It is addressed to students, who will find a basic tutorial, and all the pointers needed to become acquainted with the field; to researchers and teachers, who may use it as a reference volume; and to DA tool implementers, who need a fast way to track work on a particular problem in high-level synthesis.

One Request

The survey presented in this book will change with time. Future editions of this book will include new developments, as well as corrections — and the latter is only possible if we get feedback from the readers. We emphatically request such feedback!

Please send your comments and corrections to Robert Walker, via US Mail at the Department of Computer Science, Rensselaer Polytechnic Institute, Troy, NY, 12180, and via the Internet as "walkerb@cs.rpi.edu".

Acknowledgements

The compilation of all the papers cited in this survey would never have been possible without the kind help of many authors; we would especially like to acknowledge their efforts. We also thank Kluwer Academic Publishers for their support in publishing this survey.

Robert A. Walker
Troy, NY

Raul Camposano
Yorktown Heights, NY

Part I

Introduction to High-Level Synthesis

Introduction to High-Level Synthesis

1. Introduction

Advanced microelectronic capabilities are widely regarded as essential for an advanced industrial base. Very Large Scale Integration (VLSI) circuits are used ubiquitously in industrial products, and are critical to the progress of many fields. The competitiveness of the electronic industry requires ever-decreasing design and fabrication times for increasingly complex circuits. In addition, higher circuit speeds, close to zero defects, high yields, and application-specific integrated circuits (ASICs) produced in smaller volumes are further stressing the limits of design and fabrication. In the design of digital circuits in particular, *design automation* (DA) has been instrumental in achieving these goals.

Design automation is the *automatic synthesis* of a physical design from some higher-level behavioral specification. Automatic synthesis is much faster than manual design. It reduces the design cycle considerably, and allows the designer to experiment with various designs (design space exploration) to obtain, for example, the optimal size / speed trade-off for a given application. Furthermore, as long as the original specification was verified and simulated, a synthesized circuit should not require either verification or simulation (correctness by construction). Finally, high-level behavioral specifications are in general shorter than lower-level structural specifications (the input to logic synthesis tools), easier to write and to understand (and, therefore, change), less error-prone, and faster to simulate. Thus, they considerably facilitate the design of complex systems.

Unfortunately, design automation also exacts a price. The advantages described above are often traded for larger or slower hardware; design automation can not, in general, match the abilities of a skilled human designer. Furthermore, synthesis systems are costly and complex software; they cannot be fully verified and thus there is always the possibility for errors. Also, the theoretical advantages of correctness by construction cannot be exploited to a full extent in practice. Hence, simulation and verification of synthesized designs are a necessity. Nevertheless, the advantages of synthesis outweigh by far the problems. Today, synthesis is a growing industry, and commercial implementations of synthesis systems are widely used for production-level design of digital circuits.

Traditionally, synthesis has been subdivided into the following main categories:

- *High-level synthesis* converts a high-level, program-like specification of the behavior of a circuit into a structural design, in terms of an interconnected set of Register-Transfer (RT) level components, such as ALUs, registers, and multiplexors. With few exceptions, high-level synthesis is not yet used in practice, although this seems to be changing rapidly.
- *Logic synthesis* converts a structural design, in terms of an interconnected set of Register-Transfer level components, into optimized combinational logic, and maps that logic onto the library of available cells (in a particular technology). Two-level (PLA-like) logic synthesis systems have been in practical use for approximately one decade, and multi-level synthesis systems have recently become available.
- *Layout synthesis* converts an interconnected set of cells, which describes the structure (topology) of a design, into the exact physical geometry (layout) of the design. It involves both the placement of the cells as well as their connection (routing). Layout synthesis has been in practical use for over two decades.

An integrated synthesis system that covers all three synthesis levels is often referred to as a *silicon compiler*. In a sense, it represents the ultimate synthesis tool and the major challenge in synthesis. Such a tool would allow the design of electronic circuits from a high-level, behavioral specification with little or no human intervention.

As pointed out, logic synthesis is the highest synthesis level currently in wide-spread, practical use. The input to a logic synthesis system is a Register-Transfer Level (RTL) description — a description of an interconnected set of components, such as ALUs, adders, registers and multiplexors, with control specified using IF-THEN-ELSE and CASE-like statements. *High-level synthesis* raises the level of abstraction to the algorithmic level, allowing a more behavioral-style specification. This specification is written in a Hardware Description Language (HDL), usually a sequential (procedural, imperative) language similar to a programming language (such as C or Pascal), for example, sequential VHDL [36]. While the behavioral specification aims at describing only the functionality of a circuit — what the circuit must do, the structure gives strong hints as to the implementation — how the circuit must be built, using cells in a particular technology.

High-level synthesis bridges the gap between behavioral specifications and their hardware structure, automatically generating circuit descriptions that can be used by logic synthesis. As opposed to logic synthesis, which optimizes only combinational logic, high-level synthesis also deals with memory elements, the interconnection structure (buses and multiplexors), and the sequential aspects of a design.

High-level synthesis has been an active area of research for over 20 years. The main problems involved, such as scheduling, allocation and module assignment, have been well characterized and solved. Nevertheless, high-level synthesis systems are not yet common on the marketplace. We see four principal reasons for this.

The first reason is that high-level synthesis is a formidable task, and although the main problems have been solved, there are many issues remaining. Further optimizations involving area, cycle time, and sequential behavior (number of cycles) are necessary to obtain high quality designs. The synthesis tools should better support the specification of the interface, e.g, bus protocols, and constraints on the figures of merit of the hardware being designed, such as silicon area and timing. The control parts of a design, often specified as finite state machines, may require further optimizations, such as state minimization and encoding, and partitioning. The data path may need retiming, i.e., repositioning of memory elements to optimize timing and/or area. These are all hard problems that deserve further attention.

A second important reason is that describing hardware in terms of behavioral models may appear unnatural for certain applications. In addition, designers are often used to designing hardware at lower levels, and specifying a great deal of detail (for example, entering structural components graphically). This makes it absolutely necessary to develop synthesis systems that allow specifications at various levels. Those parts of the design that require extreme performance can thus be entered at a detailed logic level. Designers can also undertake a smooth change in their approach to design, increasingly using HDLs where it seems appropriate.

The third reason has been the lack of a generally accepted synthesis-oriented HDL in the past. Such an HDL must serve as the common input for synthesis as well as simulation, test generation, etc.

High-level synthesis research has not paid enough attention to the language aspects. Most systems start from graph representations and provide only rudimentary compilers to compile a language specification into these graphs. Although VHDL has emerged as a standard for hardware specification and simulation, the complexity of the language, the lack of a full (formal) definition of the semantics, and the impossibility of efficiently synthesizing several constructs in the language hinder its use for synthesis. This has lead to the emergence of "synthesis subsets" of VHDL, which defeat the idea of a standard language. Our hope is that as VHDL matures and undergoes revisions, most of these difficulties will be solved.

Last, the lack of effective high-level verification techniques makes the application of high-level synthesis difficult. Since correctness by construction cannot be expected from a software system with several hundred thousand lines of code, verification methods are necessary. Logic synthesis, for example, relies on Boolean verification as a feasible technique to verify that the specification and the synthesized design indeed implement the same function. Unfortunately, no effective verification techniques viable for practical-sized designs yet exist, so at present simulation is the only answer in high-level synthesis. However, verification is an active field of research and there is hope for rapid development of practical tools, though it should be pointed out that the use of unverified compiled programs is accepted practice in software development.

Despite these difficulties, high-level synthesis systems are now emerging as important tools in digital design. The rapid development of logic and finite state machine synthesis is an excellent basis for high-level synthesis, in much the same way that physical design was the basis for logic synthesis. High-level synthesis systems for specialized digital signal processing (DSP) applications have been producing practical designs for several years, and reports on practical designs produced by general purpose systems are starting to appear. Commercial systems are integrating some high-level synthesis techniques, especially allocation, into their functionality. Progress towards higher levels of design representation and more automatic synthesis will make high-level synthesis common practice for many applications in the near future.

This book presents a detailed survey of existing high-level synthesis systems. Naturally, academic approaches are in the majority, but surprisingly many industrial research systems could be included. The book begins with an introduction to high-level synthesis, which covers both design representation and the sequence of steps involved in high-level synthesis: compilation, high-level transformations, partitioning, scheduling, and allocation. The bulk of the book surveys most existing high-level synthesis systems, giving a brief description of each system's input language and internal representation, scheduling and data path synthesis algorithms, examples that have been synthesized with each system, and an extensive list of annotated references.

2. *Brief High-Level Synthesis Tutorial*

High-level synthesis is the automatic synthesis of a design structure from a behavioral specification, at levels above and including the logic level. High-level synthesis has been defined and studied systematically in [12, 46, 47]. It is usually restricted to synchronous, digital hardware synthesized from a sequential specification. For simplicity, this introduction considers only those techniques that are part of general-purpose synthesis; it does not include techniques for the synthesis of special-purpose applications such as digital signal processors or specialized architectures. Examples are kept simple, ignoring issues such as hierarchy and concurrent processes.

The main steps involved in a high-level synthesis system restricted in this way are:

- *Compilation* of the source HDL into an internal representation, usually a data flow graph and/or a control flow graph. This step is very similar to the compilation of a programming language.
- *Transformation* of the internal representation into a form more suitable for high-level synthesis. These transformations involve both compiler-like and hardware-specific transformations.
- *Scheduling*, which assigns each operation to a time step. Since only synchronous digital systems are considered here, time can be measured in so-called *control steps*, equivalent to states in a control finite state machine, or microprogram steps in a microprogrammed controller. Scheduling is sometimes called *control synthesis* or *control step scheduling*.
- *Allocation*, which assigns each operation to a piece of hardware. Allocation involves both the selection of the type and quantity of hardware modules from a library (often called module assignment) and the mapping of each operation to the selected hardware. Allocation is sometimes called *data path synthesis* or *data path allocation*.

- *Partitioning*, which divides the design into smaller pieces. Partitioning can aim at obtaining a collection of concurrent hardware modules, or can simply be used to generate smaller hardware pieces that may be easier to further synthesize and understand.
- *Output generation*, which produces the design that is passed to logic synthesis and finite state machine synthesis. The resulting design is usually generated in an RTL language.

The rest of this chapter will review the above steps in some detail.

2.1. Design Representation and Compilation

Design representations have been classified systematically by Bell and Newell [4] and Gajski and Kuhn [28], among others. Classifications commonly involve two orthogonal axes: the *level of abstraction* and the *domain*. Figure 1 shows some commonly accepted levels and domains; this is a tabular version of the well-known Y-chart [28].

Three *domains* are usually distinguished: the behavioral domain, the structural domain, and the physical domain. The *behavioral domain* describes pure behavior, ideally in terms of an input-output relationship, e.g., a set of recursive equations or a sequential program for a function, a set of Boolean equations, or a state transition diagram. The *structural domain* describes the topology of a system, typically given as a netlist of interconnected components. Components may include, for example, ALUs, multiplexors, registers, logic gates, and transistors. The *physical domain*, or layout, represents the exact geometry of the design. A geometry is given by patterns such as general polygons or rectangles, including their size and position.

For each of these representation domains, a hierarchy of *levels of abstraction* is defined, with the objects being modeled determining the level. For example, in the structural domain, the objects being connected by the netlist (i.e., the components) are processors, memories and buses at the architectural level; are registers, functional units (such as adders), multiplexors, and transfer paths at the Register-Transfer (RT) level; are latches and logic gates at the logic level; and are transistors, capacitors and resistors at the device level. This list of components is by no means exhaustive.

Level of Abstraction	Domain		
	Behavioral	**Structural**	**Physical**
Architecture	Performance, Instruction Set, Exceptions	Processors, Memory, Buses	Basic Partitions, Macrocells
Algorithmic / Register-Transfer	Algorithms, Operation Sequences	Functional Units, Registers, Connection	Floorplan
Logic (Gate)	State Transitions, Boolean Equations, Tables	Logic Gates, Latches	Cells
Device (Circuit)	Network Equations, Frequency Response, V(t), I(t)	Transistors, Capacitors, Resistors	Exact Geometry

Figure 1

Common Design Representations — Levels and Domains

It must be stressed that the boundaries of these domains and levels are not sharp. Modern logic synthesis systems, for example, accept not only Boolean equations but also arithmetic operations as their input, thus mixing the logic and the RT level. Furthermore, RT level structural languages often include conditional statements such as IF and CASE, which can be interpreted behaviorally. The list of levels given here represents only an initial approximation: the switch level, the algorithmic level, the finite state machine (cycle) level, mixed levels, etc. are also used in design automation. Furthermore, the names given here for these levels and domains are not universally accepted, adding to the confusion.

In the above representation, the starting point for high-level synthesis is a behavioral domain specification at levels above the logic level. We will assume that the behavior is specified in a sequential (procedural, imperative) Hardware Description Language (HDL) such as sequential VHDL. Synchronous systems can be described in terms of a simple model of time using basic time units called *control steps* (cycles, or states of a finite automaton). The initial design

```
entity EXAMPLE is
   port (IN1, IN2, IN3 : in integer;
         COM           : in bit;
         ...
end gcd;

architecture BEHAVIOR of EXAMPLE is
begin
     process
        variable A, B, C, D, E, F: integer;
           ...
        while (E > F)              --1
        loop
            if (COM = '1')         --2
            then
                E := IN1;          --3
                F := IN1;          --4
                A := D + IN2;      --5
                B := C - IN3;      --6
                C := A + B;        --7
                D := C + 3;        --8
            else
                E := IN2;          --9
                F := IN3;          --10
            end if;                --11
        end loop;
        ...
```

Figure 2

A Behavioral Specification Fragment in VHDL

specification for high-level synthesis typically includes only a partial timing specification. Furthermore, the limitation that we consider only digital systems restricts the modeled operations to well-known arithmetic and logical operations.

Figure 2 shows an example of a behavioral specification. It contains a simple VHDL program fragment, demonstrating assignment, arithmetic operations, a WHILE loop, and an IF...THEN...ELSE statement.

Compilation transforms the design representation, written in an HDL, into an internal representation more useful for high-level synthesis. As in the compilation of programming languages, the intermediate representations are usually graphs and parse trees, although the vast majority of high-level synthesis approaches use graphs. Examples of such representations are the Yorktown

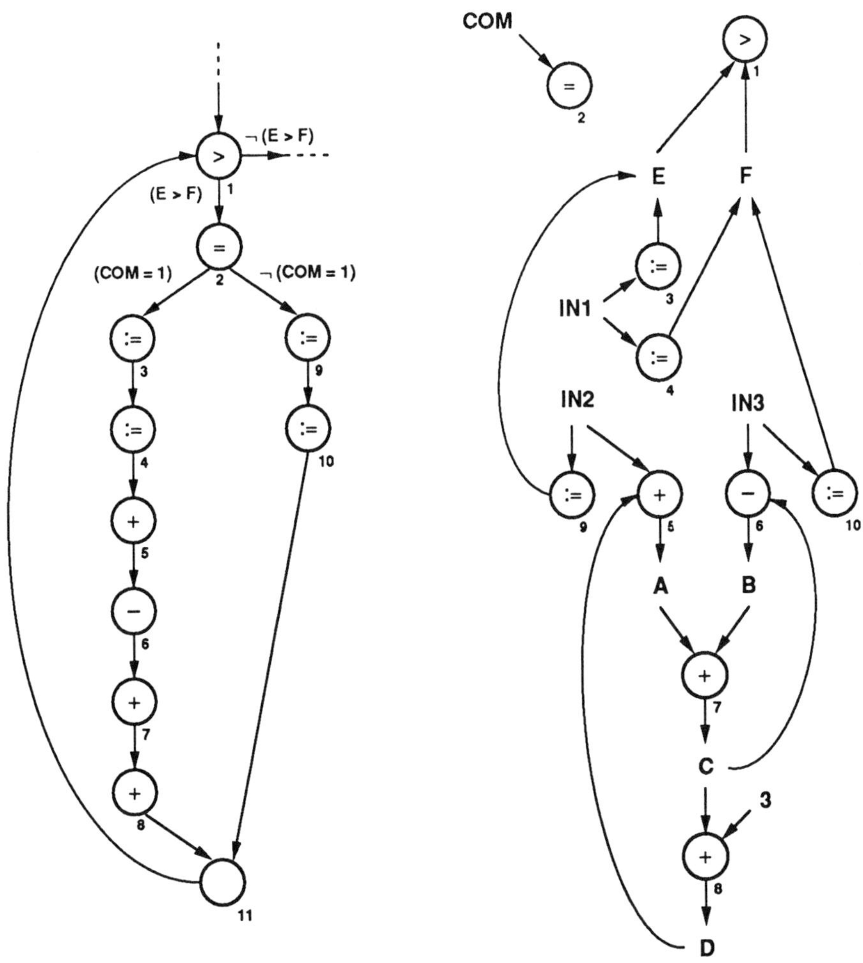

(a) Control Flow Graph **(b) Data Flow Graph**

Figure 3

Complete Data Flow and Control Flow for EXAMPLE

Intermediate Format (YIF) [10], IBM's Sequential Synthesis In-core Model (SSIM) [19], CMU's well-known Value Trace (VT) [43], USC's Design Data Structure (DDS) [37], Irvine's Behavioral Intermediate Format [50] and Karlsruhe's Internal Format (IF) [17].

Compilation is basically a one-to-one transformation from the behavioral specification into the internal representation, with optimization done in a second step (described in the next section).

Continuing the previous example, separate data-flow and control-flow graphs are generated, as shown in Figure 3. Nodes are numbered with the labels provided in comments in the VHDL model. Operation nodes in the data-flow and control-flow graphs correspond to each other, and are labeled with the operation they represent, e.g., "+". The reader can easily verify the one-to-one correspondence to the specification of EXAMPLE in Figure 2.

The *control-flow graph* (CFG) is directed, and indicates the control flow in the original specification. The nodes correspond to the operations, and the edges link immediate predecessor-successor pairs. For example, operation 8 follows (is an immediate successor of) operation 7. The CFG also indicates loops, for example, edge (11, 1) closes the WHILE loop. Conditional branching is indicated by more than one successor, e.g., operation 2 has 2 successors. Exactly one successor will be selected, depending on the condition attached to the edge. For example, operation 3 will follow operation 2 if (COM=1), i.e., if the result of the preceding comparison, operation 2, is true.

The *data-flow graph* (DFG) is also a directed graph. The nodes represent the data (variables and signals) and the operations (shown as circles). The directed edges indicate the direction of the data flow. For example, operation 5 (an addition) reads IN2 and D as inputs, and writes A as its output. The given graph represents the complete data flow; in later sections it will be shown that for certain applications it can be combined with the CFG and significantly simplified. Note that the DFG may be disconnected, as shown by operator 2.

The method described here of separating the data and control flow is only one of a variety of ways that high-level synthesis systems represent this information. Some systems choose to represent the operations only once, mixing both graphs. Other systems preserve only the essential parts of the control flow, such as loops and conditional branches. For example, the ordering of operations 5 and 6 is not relevant, since they are not data dependent, so that ordering does not have to be stored. Unfortunately, such mixed data/control flow representations are also occasionally called "data-flow" representations, which creates some confusion. For didactical purposes, we have chosen to separate data flow and control flow in this book.

Figure 4 shows a simplified *dependency graph* for EXAMPLE, formed by combining and simplifying the DFG and CFG. It contains only the essential operation dependencies, i.e., data dependencies (such as between operations 5 and 7), resource conflicts (operations 3 and 4

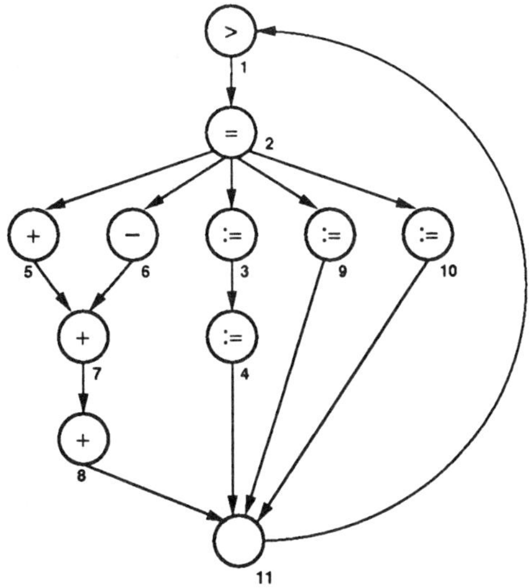

Figure 4
Simplified Dependency Graph for EXAMPLE

both use IN1, which is assumed to be an input port, so they must read the values sequentially), and some control dependencies (operation 5 can only be executed after the conditional branch operation 2). Such a simplified representation can be of great value for scheduling. It does not contain, however, the complete set of information contained in the original DFG and CFG. Operations 3 and 9, for example, are mutually exclusive (they can never be executed in parallel because they are in different conditional branches), while operations 6 and 3 are not mutually exclusive; this organization is represented in the CFG, but not in the dependency graph.

2.2. Transformations

High-level *transformations* aim at optimizing the behavior of the design. Obvious transformations are compiler optimizations [1], such as constant propagation, dead code elimination, common subexpression elimination, variable disambiguation by global flow analysis, code motion, in-line expansion of subprograms, and loop unrolling. Other transformations are more specific to high-level synthesis, including:

- Hardware-specific local transformations, for example, substituting multiplication by a power of two with the selection of the appropriate bits (equivalent to a shift), taking into account commutativity of operators, etc. [59].

- Increasing operator-level parallelism — in principle, an arbitrary number of operations can be performed in parallel, provided there is sufficient hardware.
- Reducing the number of levels in the data-flow or control-flow graphs, such as in FLAMEL [63], which may lead to faster hardware.
- Creating concurrent processes, which may be synthesized as concurrent hardware, thus increasing the speed of the design [60, 67]. A special case of concurrent process creation is pipeline creation.

In practice, it is difficult to evaluate the exact effect of high-level transformations. At this point in the synthesis process, measures such as size, power, and delay are only estimates, as they can be evaluated exactly only when the real hardware is generated. For example, what may seem to be a critical path with respect to timing may turn out to be faster than other paths after logic synthesis. In contrast, a path that appears to be fast at a high level may turn out to be critical because of unexpected wire delays. Thus, high-level transformations must be applied with care, and often only the designer can decide which transformations to apply.

As an example of transformation, Figure 5 shows a partial loop unrolling applied to EXAMPLE. The simplified dependency graph (Figure 4) was duplicated twice, and only the real dependencies are shown. Operation 1 of the second iteration does not depend on the result D of operation 8, which is only used by operation 5. Similarly, only operation 6 uses the result C of operation 7. Thus, operations 7 and 8 can be executed in parallel with the test at the beginning of the next loop iteration (operation 1). The partially-unrolled dependency graph captures this information.

2.3. Scheduling

Scheduling assigns each operation in the behavior to a point in time; in synchronous systems, time is measured in *control steps*. Scheduling aims at optimizing the number of control steps needed for completion of a function, given certain limits on hardware resources and cycle time. A scheduling algorithm must take into account the control constructs, such as loops and conditional branching, the data dependencies expressed in the data-flow graph, and constraints on the hardware. In synchronous hardware the basic constraints are that every unit of hardware can be used only once during a control step, i.e., registers can

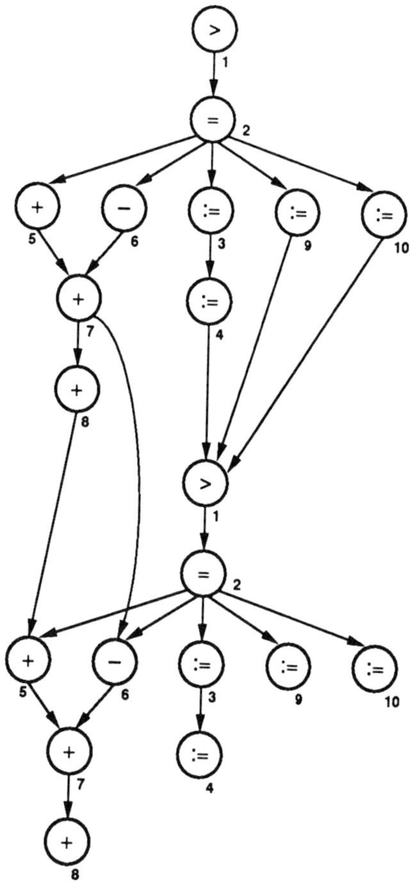

Figure 5

EXAMPLE After Partial Loop Unrolling

be loaded only once, combinational logic may evaluate only once (feedback is forbidden), and buses may carry only one value. Other constraints on a design may restrict the size, the delay and the power.

The first approach to scheduling in high-level synthesis was probably the exhaustive search in Expl [3]. Since then, many scheduling algorithms for high-level synthesis have been proposed, some relying on methods known from microprogram optimization. Davidson et. al. [25] discuss exhaustive search using branch-and-bound techniques. In addition, they study First-Come-First-Served

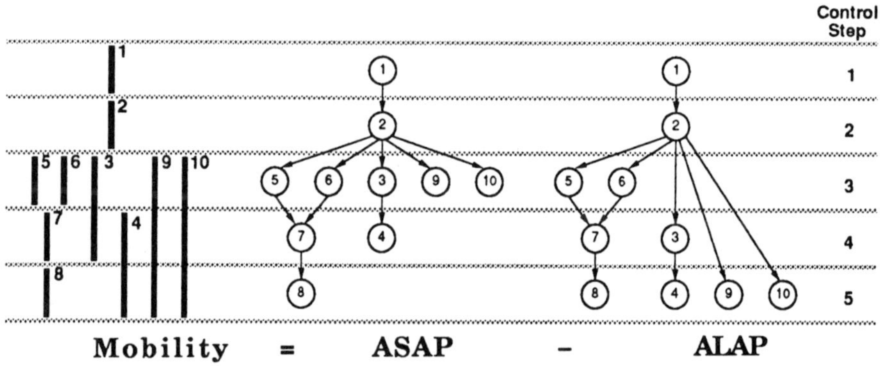

Figure 6

Mobility, ASAP Schedule, and ALAP Schedule for EXAMPLE

(FCFS, also known as As-Soon-As-Possible, or ASAP) scheduling, list scheduling, and scheduling the critical path first.

Figure 6 uses the dependency graph from Figure 4 to explain some of these scheduling concepts. In the As-Soon-As-Possible (ASAP) schedule, all operations are assigned to the earliest possible control step, corresponding to a topological sort of the graph in depth-first order. The As-Late-As-Possible (ALAP) schedule assigns all operations to the latest possible control step. The *mobility*, then, is the difference of the ASAP and ALAP schedules. Only 5 control steps were used, so all operations on the critical path have a mobility of one (they have to be scheduled into that one possible control step). Operation 3, however, has a mobility of two, and can be scheduled in either control step 3 or 4. It should be noted that this simple model assumes that each operation uses exactly one control step. *Chaining* (scheduling two or more dependent operations in one control step) and *multi-step operations* (operations which need two or more control steps to complete) were not considered in this example.

Another scheduling approach is *list scheduling*, which schedules operations into control steps, one control step at a time. For the current control step, a list of *data ready* operators is constructed, containing those operators whose inputs are produced in earlier control steps, and that do not violate any resource constraints. This list is then sorted according to some *priority function*, the highest-priority operator is placed into the current control step, the list is updated, and the process continues until no more operators can be placed into that control step. This process is then repeated on the next control step, until the entire design is scheduled. Two common priority functions are mobility

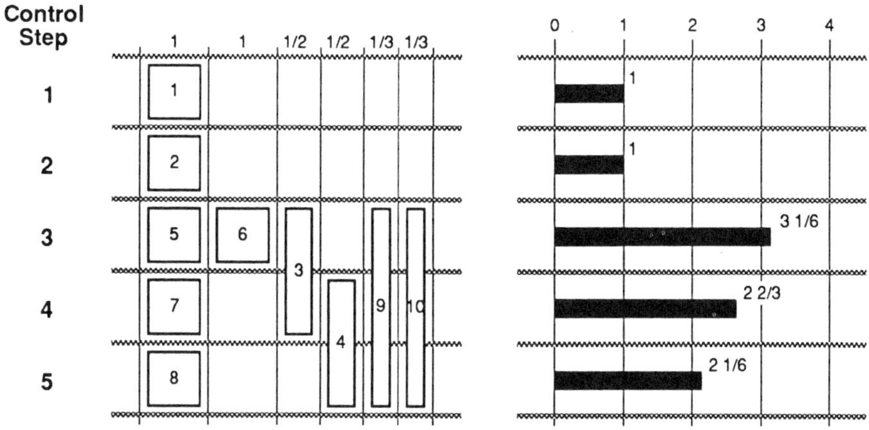

(a) Time Frames **(b) Distribution Graph**

Figure 7

Time Frames and Distribution Graph for EXAMPLE

(schedule the operator with the smallest mobility first) and path length (schedule the operator on the longer path first). For the example, the priority function would choose the operators in the order 1, 2, 5, 6, 3, 9, 10, 7, 4, 8, using either mobility or path length as the priority function. If we limit the number of operations to three per control step, operations 5, 6 and 3 will be scheduled in step 3, operations 7, 9 and 10 in step 4, and operations 8 and 4 in step 5.

Variations of list scheduling are used in many high-level synthesis systems. CMU's System Architect's Workbench [69] uses list scheduling that considers timing constraints, resource constraints, and resource utilization in its priority function, AT&T's Algorithms to Silicon Project uses list scheduling with the path length to the end of the graph as the priority function [44], MAHA [53] and SLICER [51] use list scheduling with mobility (or freedom) as the priority function, and Karlsruhe's system [39] uses list scheduling attempting to minimize delay.

A more complex scheduling method is *force-directed scheduling* [55], used in HAL [54] and SAM [23]. Force-directed scheduling aims at balancing the number of operations in each control step. To do this, the mobility is first quantified into *time frames* (Figure 7(a)). The width of a time frame is the probability of an operation being scheduled in that particular control step, initially 1/n, where n is the number of possible control steps in which that operation can be scheduled. For example, operation 1 must be scheduled in control step 1 and thus has a time frame

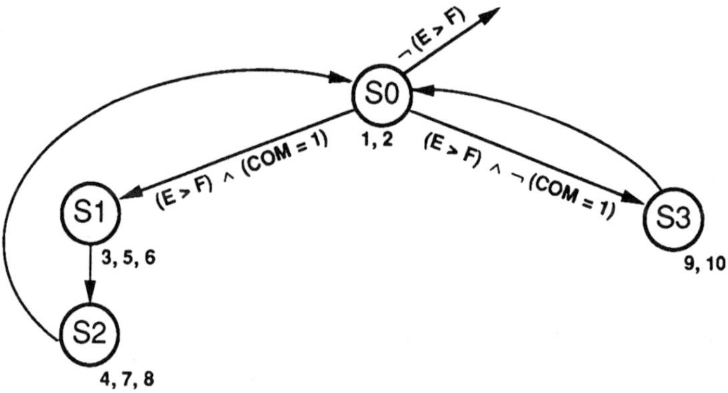

Figure 8
FSM for EXAMPLE
Restricted to 2 ALUs, with Chaining of 2 Operations Allowed

of width 1. Operation 9 can be scheduled in control steps 3, 4 or 5; it has a probability of 1/3 to be in each of these control steps.

Adding the time frames up over each control step, the *distribution graph* is obtained (Figure 7(b)). This graph measures how crowded each control step is, i.e., how many operations will probably be executed in the control step, hence, how much hardware will be required. Operations are scheduled so that the distribution graph has similar values for all control steps. The effect for each possible assignment of an operation to a control step is calculated, scheduling the operation / control step pair that minimizes the distribution differences among control steps. Fixing an operation to a particular control step changes the time frame for the operation. For example, scheduling operation 9 into control step 5 will change its time frame to just control step 5, thus increasing the distribution graph for step 5 by 2/3 and decreasing the time frames for control steps 3 and 4 by 1/3. The quantitative effect of scheduling each operation to each possible control step is formulated with equations similar to force equations, hence the name. The scheduling choice with the largest negative force (the largest smoothing effect on the distribution graph) is selected, and the whole process of computing the distribution graph and forces is then repeated until all operations are scheduled.

The example in Figure 4 can be scheduled as indicated by the finite state machine (FSM) diagram shown in Figure 8 (recall that in an FSM, the states correspond to control steps). Both list scheduling and force-directed scheduling will produce this result, if we limit the number of arithmetic operations (+ and –) to 2 per control step, and allow chaining of up to two operations per control step. Notice that operations 9 and 10 are part of the ELSE branch (Figures 2 and 3), and must be scheduled in a separate control step (state) from the operations in the IF branch. Also, operations 3 and 4 can not share the same control step, because they both use the same port IN1 (only one value per pin per control step can be read in a synchronous design). The resulting finite state machine has four states. The state transition conditions can be derived directly from the original specification.

The above scheduling algorithms are *constructive* in the sense that they construct a schedule from the specification. Another group of scheduling algorithms *transforms* a given schedule into another one. In trace scheduling [27] and percolation scheduling [49], operations are moved to earlier control steps, possibly vacating control steps, which can then be eliminated. Operations can be moved only if their dependencies and the constraints allow the move. The V system [6], the DSL/CADDY system [17], CAMAD [56] and VSS [57] all use variations of this method. Other transformational techniques that have been used for scheduling are simulated annealing [26], self-organization (neural nets) [34] and control state splitting (the opposite of merging such as in percolation scheduling) [13].

We finally review scheduling methods which put more emphasis on conditional branching. The main idea of these algorithms is to schedule mutually exclusive operations to allow sharing of hardware, and possibly a faster schedule for some paths in the control flow graph.

Path-based scheduling [11] is one of these scheduling methods. It first computes all paths in the control flow graph, then schedules each path independently, selecting the particular order of the operations on each path, for example, using list scheduling. Figure 9 gives the two paths for the example in Figure 3. In this example, the second path can be scheduled in just one control step. The first path, however, has constraints, as indicated as the bars in Figure 8, which mean that a new control step must start at the beginning of some operation covered by the bar. The first constraint covering operations 7, 8, and 4 arises because operations 3 and 4 use the same resource (read from IN1), hence a new control step must start at operation 7, 8 or 4 to avoid scheduling operations 3 and 4 in the same control step. The other two constraints come from limiting the number of arithmetic operations to 2 (i.e., operations 5, 6 and 7 cannot be scheduled into the same control step, nor can 6, 7 and 8). The final schedule is built by determining the best schedule for each path, taking the constraints into account. In the

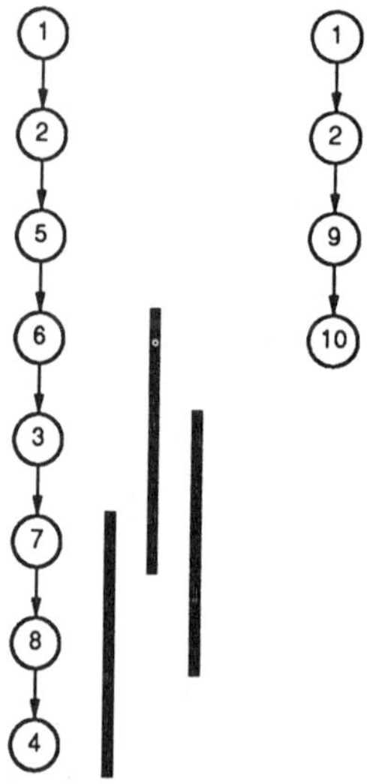

Figure 9
Paths and Constraints for EXAMPLE

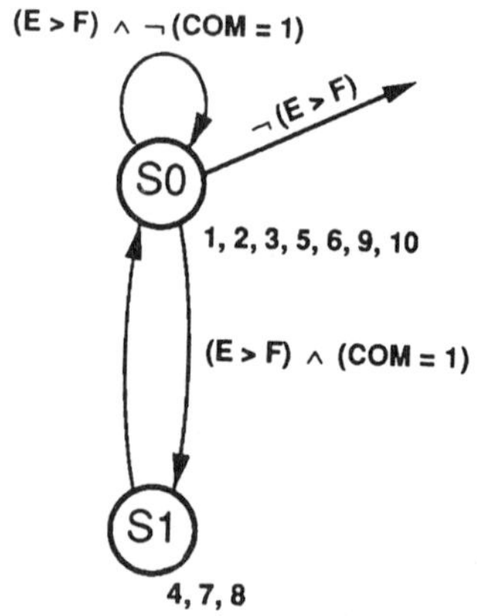

Figure 10
FSM for EXAMPLE, Based on AFAP Scheduling

example, two states suffice for path one, i.e., the second state starts at operation 7. Finally, the states are merged whenever their first operation is the same, resulting in the finite state machine shown in Figure 10.

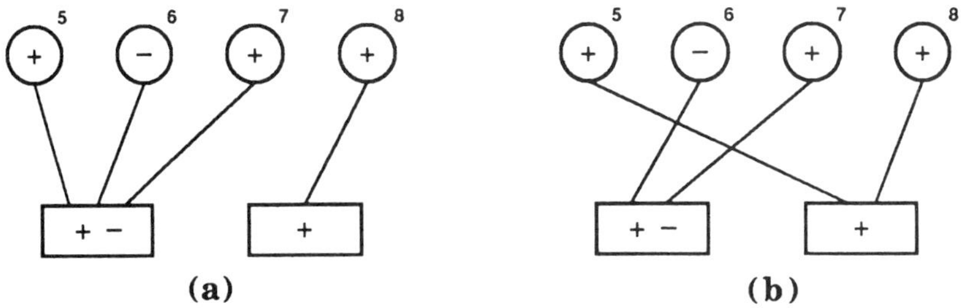

Figure 11
Module Assignment for EXAMPLE

Path-based AFAP scheduling [11] formulates the above method as a clique covering problem; a similar technique was used in the YSC [9]. Other scheduling approaches that consider conditional branching are Bridge's global slicing [65] and resource sharing [66].

Although scheduling is NP-complete in general, today it is well-understood. Many heuristics produce very good results, and can be applied to large, practical problems.

2.4. Allocation

Allocation assigns each operation, variable, and communication path to a piece of hardware. It naturally falls into three parts: *functional unit (FU) allocation*, *register allocation* and *connection allocation*. In high-level synthesis, the main aim in allocation is to share hardware units, i.e., operations can share functional units (ALUs, adders, etc.), variables can be mapped onto common registers, and memories and transfers can share buses and multiplexors. The goal of allocation is to optimize the overall hardware.

A problem related very closely to allocation is *module assignment*. Whenever there is more than one type of functional unit to perform a particular operation, that operation has to be assigned to one specific functional unit type. Consider the two possible module assignments for the arithmetic operations of the example (Figure 4) given in Figure 11, assuming that the two FUs (+, –) and (+) are available. If operation 5 is assigned to the FU (+, –), since operation 6 has to be assigned to (+, –), these two operations can not be scheduled in parallel (Figure 11(a)), and 3 control steps would be required to schedule all four operations. However, if alternative (b) is picked, operations 5 and 6 can be scheduled in the same control step and all four arithmetic operations require only 2 control steps. A heuristic to solve this problem is given in [5].

Three main types of allocation algorithms can be distinguished:

- Heuristic allocation, e.g., greedy and sequential allocation.
- Linear programming approaches.
- Graph-based algorithms.

Heuristic approaches usually select one element (operation or variable) at a time to allocate and assign to hardware, with the selection done according to different criteria. EMUCS [35, 62] used a global cost function; the assignment that saves the most hardware is the one selected. Another approach is to use the data-flow order [32]. Expert systems, such as the Design Automation Assistant [38], and systems that use graph grammars, such as Elf [30], usually apply the rule that results in the lowest cost; these can also be considered heuristic approaches. Heuristic approaches yield reasonable results, are fast, and have the potential to mix FU, register and communication path allocation.

Linear programming (LP) approaches formulate allocation (or allocation and scheduling) as a linear programming problem (0-1 LP, mixed integer LP, integer programming), and then solve that problem. This approach was pioneered in high-level synthesis by Hafer and Parker [33]. In the past, however, since linear programming required extensive computational resources, it could only be used for small examples. Recently, the method has reappeared, among other reasons because modern large LP systems can solve problems with tens of thousands of variables. Examples are Tsing Hua University's system [42], allocation of multi-port memories [2], and integer programming in [29].

The third group of algorithms formulate allocation as *clique covering* (or *partitioning*) of a compatibility graph, or *node coloring* of the dual (conflict) graph. In high-level synthesis, Tseng and Siewiorek first formulated allocation as a clique covering problem [64]. An equivalent problem to clique covering is coloring the dual graph, which has been used in compiler construction for a long time, e.g., [22]. Figure 12 illustrates node coloring (a) and clique covering (b), allocating functional units for the simplified dependency graph in Figure 4 with the schedule from Figure 10.

The graph shown in Figure 12(a) is called a *conflict graph* — nodes represent operations, and edges represent conflicts. For example,

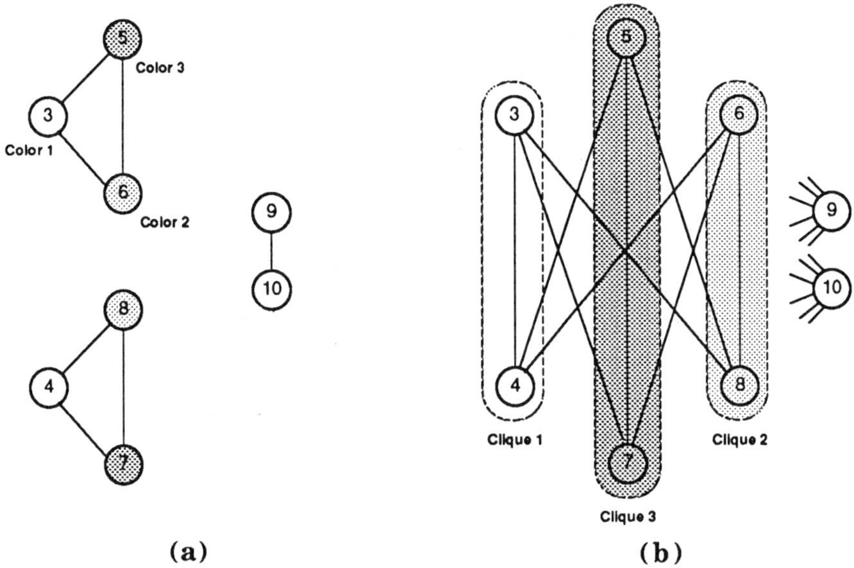

Figure 12
Allocation Using Conflict and Compatibility Graphs

operations 3, 5 and 6 conflict, because they are scheduled in the same control step, and thus cannot be assigned to the same functional unit. Operations 9 and 10 have no conflicts with operations 3, 5 and 6, even though they are scheduled in the same control step, because the two branches are mutually exclusive (Figure 3). With node coloring, the number of colors needed to color the nodes, such that no two adjacent nodes will have the same color, is the minimum number of functional units needed. In Figure 12(a), three colors suffice. However, allocation may choose not to share colors for operations 9 and 10 since they are simple data transfers and cost nothing from the point of view of functional units.

The graph shown in Figure 12(b) is called a *compatibility graph*, and is the dual of the conflict graph. For simplicity in the figure, the edges from operations 9 and 10 are not shown in their entirety. Cliques (totally connected subgraphs) in the compatibility graph represent sets of operations which can share hardware; hence, a minimum *clique covering* gives the minimum number of functional units. In clique covering, a cost is often added to the edges to indicate the cost of sharing hardware by the two adjacent nodes. In this case, a minimum cost clique cover represents the best solution.

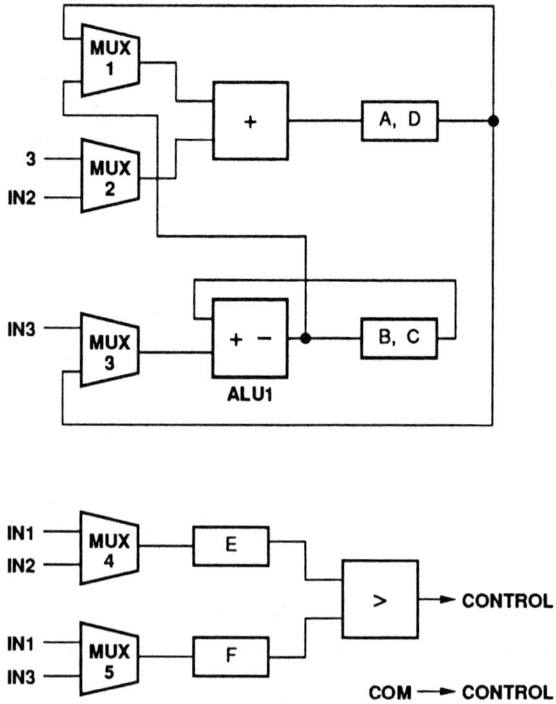

Figure 13
Data Path for EXAMPLE, Using the ASAP Schedule

The register allocation problem and the communication allocation problem can be formulated in a similar way. Variables with non-overlapping lifetimes are compatible, as are mutually exclusive data transfers.

In general, the node coloring and clique covering problems are NP complete. However, in special cases, the problems may be simplified. In the REAL system, register allocation for a single basic block (no loops or conditional branches) can be done by a simple left edge algorithm [40]. In the System Architect's Workbench, Springer and Thomas have shown that certain types of graphs can be colored in polynomial time [61].

Using two functional units for the arithmetic operations as obtained in Figure 12 (colors 2 and 3), and allocating registers and communication (multiplexors and wires) accordingly, the data path of

Inputs			**Next State**	**Outputs**									
STATE	**E>F**	**COM =1**		**ALU 1**	**A, D**	**B, C**	**E**	**F**	**MX 1**	**MX 2**	**MX 3**	**MX 4**	**MX 5**
State0	1	0	State0	x	0	0	1	1	x	x	x	1	1
State0	1	1	State1	–	1	1	1	x	0	1	0	0	x
State0	0	x		x	x	x	x	x	x	x	x	x	x
State1	x	x	State0	+	1	1	0	1	1	0	1	x	0

Signal Specification: x = don't care; ALU1: + selects add, – selects sub; Registers: 0 = keep, 1 = load; MUXes: 0 = top, 1 = bottom

Figure 14
Control Specification for EXAMPLE (Figures 10 & 13)

Figure 13 is allocated. Notice that the connection from ALU1 through MUX1 to the adder allows the chaining of operations 7 and 8.

To complete the design, control hardware must be built. At this point, the control is already totally specified, as the finite state machine (or more precisely, the states and state transition logic) was obtained during scheduling. The required output signals can be determined after data-path allocation, i.e., which function the FUs are to perform in each control step, which registers to load, and which inputs to select at each multiplexor. Control for one sequential design can be implemented by one finite state machine, which can be synthesized by standard techniques using state assignment, variable encoding and logic synthesis. The finite state machine for the example (Figures 10 and 13) is given in Figure 14.

The scheduling and allocation tasks are highly interdependent, and it is not clear in which order they should be performed. Most approaches schedule first, relying on estimations of the required amount of hardware. Some approaches allocate the data-path first, subject to given constraints, and then schedule taking into account the already given hardware. Yet other approaches intertwine scheduling and allocation, attempting to optimize a cost function that includes both the number of control steps and the hardware. The ultimate criterion for evaluating synthesized hardware is the design by an expert human designer, who may use design skills very unlikely to be emulated by an automated system.

2.5. *Partitioning and Output Generation*

For large designs, it is desirable to *partition* the design into smaller pieces that can be processed more or less independently. This is often referred to as hierarchical design. Different high-level synthesis systems partition the design at different stages. Some assume that a design is partitioned into modules by the initial specification, and process these modules separately [9, 14]. Others partition a specification early in the synthesis process: BUD [44, 45] partitions the design using clustering, APARTY [41] partitions an instruction set using multistage clustering, and Stanford's Olympus system [31] uses simulated annealing and the Kernighan-Lin algorithm to partition functional models. The YSC also partitions a design using clustering [15], but at the end of the high-level synthesis process, just before the design is passed to logic synthesis.

The result of high-level synthesis is often passed to *logic synthesis* tools, which optimize the combinational logic, and then to *layout synthesis* tools, which design the chip geometry. *Output generation* must generate the design in a format that these tools can process it. In an integrated design system, the design is stored in a data base and all programs can access it directly. Stand-alone tools generate a specification of the design in a low level hardware description language, for example, structural VHDL or EDIF.

2.6. *Open Problems*

Concurrency is still one of the most important issues that need to be further investigated. Concurrency at the operation level is handled adequately by scheduling, and concurrency at lower levels, e.g., the logic level, is addressed by logic synthesis, but at higher levels concurrency remains a major concern. The automatic synthesis of pipelines, for example, has only been addressed for special cases such as pipelines with no flushing, restart and stage bypasses, e.g., SEHWA [52], or for specialized digital signal processing applications, e.g., PISYN [21]. Systems like CATHEDRAL-II [58] automatically synthesize a predefined architecture of concurrent processors, but specialized for DSP applications. Transformations that obtain concurrency for more general kinds of applications at the system level have been defined at CMU [68], but must be applied interactively. Thus, there is no general understanding on how to obtain concurrency automatically at the system level, nor are there general purpose systems of this kind in practical use.

A more specialized topic that is only recently being addressed in detail is synthesis with constraints such as delay, area and power. Most scheduling methods allow the consideration of area and / or delay constraints, e.g., force-directed scheduling, path-based scheduling, and list scheduling. However, timing constraints in general, and synthesis with interfaces (e.g., a bus requiring a given protocol), are more difficult problems, and have been studied by Camposano and Kunzmann [16], Nestor and Thomas [48] and Borriello [7].

Despite the large amount of work and despite standardization efforts, the use of HDLs for synthesis is still an unresolved issue. Although VHDL [36] has emerged as a standard for hardware specification, some questions remain as to its applicability as an input language for high-level synthesis. For example, it is unclear what class of applications should be described in a sequential or a functional style, for behavioral specification and synthesis. Furthermore, several language constructs, well suited for simulation, are difficult to synthesize (e.g., dynamic memory allocation and sensitivity lists [18]), or are not applicable for synthesis (e.g. assertions).

The formal foundation of high-level synthesis is also still in its infancy. This state is partially due to the lack of formal semantics for hardware description languages, such as VHDL, as well as the imprecise formulation of some synthesis problems, usually due to the number of assumptions being made (e.g., the type of registers, the clocking scheme, and the implementation of central / distributed multiplexing). The absence of a coherent theory that would allow a practical way of proving the equivalence of a behavioral and a structural description at levels higher than the logic level is also a problem, although current developments in formal verification may help in this area. Verification of finite state machines, for example, has made impressive advances lately [24].

Last, but not least, several design process issues deserve more attention.

- The user interface, manual design intervention, and the role of graphics should be studied further.

- When a process is decomposed into independent steps, i.e., high-level synthesis on one side and logic synthesis or layout on the other, problems arise with the flow of information in both directions [20]. High-level synthesis must pass detailed design constraints, not only structural information, to logic synthesis. High-level synthesis must receive lower-level implementation information such as size, delay and power from logic synthesis.

- Structural issues, such as the advantage of regularity, the level of the primitives, and the influence of the size and the functionality of the partitions of combinational logic generated must be considered in high-level synthesis. Non-structural information, such as the underlying control scheme, constraints on time and power, the required testability, and the required technology libraries, should also be considered.
- The considerable gap between high-level synthesis and logic synthesis must be bridged, although the synthesis of finite state machines, involving state encoding, encoding of symbolic variables in logic synthesis, and retiming techniques, are signs that this bridge is beginning [8].
- A strict top-down or meet-in-the-middle approach makes it rather difficult to meet stringent constraints on size, delay, etc. Design iteration in a "closed loop" may help, requiring much tighter system integration.

In spite of these open questions, the body of knowledge in high-level synthesis has reached a state in which practical applications seem possible. The many systems reviewed in the rest of the book are a clear indication of the amount of work already done.

References

[1] A.V. Aho, R. Sethi, and J.D. Ullman, *Compilers: Principles, Techniques and Tools*, Reading, MA: Addison-Wesley, 1986.

[2] M. Balakrishnan, A.K. Majumdar, D.K. Banerji, and J.G. Linders, "Allocation of Multi-Port Memories in Data Path Synthesis," *IEEE Transactions on Computer-Aided Design*, vol. 7, no. 4, pp. 536-540, April 1988.

[3] M. Barbacci, *Automatic Exploration of the Design Space for Register Transfer (RT) Systems*, PhD thesis, Department of Computer Science, Carnegie-Mellon University, November 1973.

[4] C.G. Bell and A. Newell, *Computer Structures: Readings and Examples*, New York: McGraw-Hill, 1971.

[5] R. Bergamaschi, R. Camposano, and M. Payer, "Area and Performance Optimizations in Path-Based Scheduling," *Proc. of EDAC'91*, Amsterdam, February 1991.

[6] V. Berstis, "The V Compiler: Automatic Hardware Design," *IEEE Design & Test of Computers*, pp. 8-17, April 1989.

[7] G. Borriello, "Combining Event and Data-Flow Graphs in Behavioral Synthesis," *Proceedings of the ICCAD'88*, pp. 56-59, Santa Clara, Ca, November 1988.

[8] G. Borriello and A. Sangiovanni-Vincentelli, "Special Issue on HICCS'89 - Sequential Logic Synthesis," *IEEE Transactions on Computer-Aided Design*, vol. 10, no. 1, January 1991.

[9] R.K. Brayton, R. Camposano, G. DeMicheli, R.H.J.M. Otten, and J.T.J. van Eijndhoven, "The Yorktown Silicon Compiler System," in D. Gajski, editor, *Silicon Compilation*, Addison-Wesley, 1988.

[10] R. Camposano, "Structural Synthesis in the Yorktown Silicon Compiler," in C.H. Sequin, editor, *VLSI'87, VLSI Design of Digital Systems*, pp. 61-72, Vancouver: North-Holland, 1988.

[11] R. Camposano, "Path-Based Scheduling for Synthesis," *Proceedings of the HICSS'90*, pp. 348-355, Kailua-Kona, Hawaii, January 1990.

[12] R. Camposano, "From Behavior to Structure: High-Level Synthesis," *IEEE Design and Test of Computers*, vol. 7, no. 5, pp. 8-19, October 1990.

[13] R. Camposano and R.A. Bergamaschi, "Redesign Using State Splitting," *Proceedings of the First EDAC*, pp. 157-161, Glasgow, UK, March 1990.

[14] R. Camposano, R. Bergamaschi, C. Haynes, M. Payer, and S. Wu, "The IBM High-Level Synthesis System," in R. Camposano and Wayne Wolf, *Trends in High-Level Synthesis*, Norwell, MA: Kluwer Academic Publishers, 1991.

[15] R. Camposano and R.K. Brayton, "Partitioning Before Logic Synthesis," *Proceedings of the ICCAD'87*, pp. 237-246, Santa Clara, CA, November 1987.

[16] R. Camposano and A. Kunzmann, "Considering Timing Constraints in Synthesis from a Behavioral Description," *Proceedings of the ICCD'86*, pp. 6-9, Port Chester, October 1986.

[17] R. Camposano and W. Rosenstiel, "Synthesizing Circuits from Behavioral Descriptions," *IEEE Transactions on Computer-Aided Design*, vol. 8, no. 2, pp. 171-180, February 1989.

[18] R. Camposano, L.F. Saunders, and R.M. Tabet, "High-Level Synthesis from VHDL," *IEEE Design and Test of Computers*, March, 1991.

[19] R. Camposano and R.M. Tabet, "Design Representation for the Synthesis of Behavioral VHDL Models," in J.A. Darringer, F.J. Rammig, editors, *Proceedings CHDL'89*, Elsevier Science Publishers, 1989.

[20] R. Camposano and L.H. Trevillyan, "The Integration of Logic Synthesis and High-Level Synthesis," *Proceedings ISCAS'89*, IEEE, 1989.

[21] A.E. Casavant, M.A. d'Abreu, M. Dragomirecky, D.A. Duff, J.R. Jasica, M.J. Hartman, K.S. Hwang, and W.D. Smith, "A Synthesis Environment for Designing DSP Systems," *IEEE Design and Test of Computers*, pp. 35-45, April, 1989.

[22] G.J. Chaitin, M.A. Auslander, A.K. Chandra, J. Cocke, M.E. Hopkins, and P.W. Markstein, "Register Allocation via Coloring," *Journal on Computer Languages*, vol. 6, pp. 47-57, 1981.

[23] R. Cloutier and D. Thomas, "The Combination of Scheduling, Allocation and Mapping in a Single Algorithm," *Proceedings of the 27th Design Automation Conference*, pp. 71-76, Orlando, FL, June 1990.

[24] O. Coudert, C. Berthet, and J.C. Madre, "Formal Boolean Manipulations for the Verification of Sequential Machines," *Proceedings EDAC'90*, pp. 57-61, Glasgow, Scotland, March 1990.

[25] S. Davidson, D. Landskov, B.D. Shriver, and P.W. Mallet, "Some Experiments in Local Microcode Compaction for Horizontal Machines," *IEEE Transactions on Computers*, vol. C-30, no. 7, pp. 460-477, July 1981.

[26] S. Devadas and A.R. Newton, "Algorithms for Allocation in Data Path Synthesis," *IEEE Transactions on Computer-Aided Design*, vol. 8, no. 7, pp. 768-781, July 1989.

[27] J.A. Fisher, "Trace Scheduling: A Technique for Global Microcode Compaction," *IEEE Transactions on Computers*, vol. C-30, no. 7, July 1981.

[28] D. Gajski and R. Kuhn, "Guest Editors' Introduction: New VLSI tools," *IEEE Computer*, vol. 16, no. 12, pp. 11-14, December 1983.

[29] C.H. Gebotys and M.I. Elmasry, "A Global Optimization Approach for Architectural Synthesis," *Proceedings ICCAD'90*, pp. 258-261, Santa Clara, CA, November 1990.

[30] E.F. Girczyc and J.P. Knight, "An ADA to Standard Cell Hardware Compiler based on Graph Grammars and Scheduling," *Proc. ICCD'84*, October 1984.

[31] R. Gupta and G. DeMicheli, "Partitioning of Functional Models of Synchronous Digital Systems," *Proceedings ICCAD'90*, pp. 216-219, Santa Clara, CA, November 1990.

[32] L.J. Hafer and A.C. Parker, "Register-Transfer Level Design Automation: The Allocation Process," Proc. *15th Design Automation Conference*, pp. 213-219, June 1978.

[33] L.J. Hafer and A.C. Parker, "A Formal Method for the Specification, Analysis and Design of Register-Transfer Level Digital Logic," *IEEE Transactions on CAD*, vol. 2, no. 1, pp. 4-18, January 1983.

[34] A. Hemani and A. Postula, "A Neural Net Based Self Organising Scheduling Algorithm," *Proceedings of the EDAC'90*, pp. 136-140, Glasgow, Scotland, March 1990.

[35] C.Y. Hitchcock III and D.E. Thomas, "A Method of Automated Data Path Synthesis," *Proceedings 20th Design Automation Conference*, pp. 484-489, June 1983.

[36] IEEE, *Standard VHDL Language Reference Manual*, New York: The Institute of Electrical and Electronics Engineers, Inc., March 1988.

[37] D.W. Knapp and A.C. Parker, "A Unified Representation for Design Information," *7th International Symposium on Computer Hardware Description Languages and their Applications*, pp. 337-353, Tokyo, August 1985.

[38] T.J. Kowalski, *An Artificial Intelligence Approach to VLSI Design*, Boston: Kluwer Academic Publishers, 1985.

[39] H. Kramer and W. Rosenstiel, "System Synthesis using Behavioral Descriptions," *Proceedings of the EDAC'90*, pp. 277-282, Glasgow, Scotland, March 1990.

[40] F.J. Kurdahi and A.C. Parker, "REAL: A Program for Register Allocation," *Proc. 24th Design Automation Conference*, ACM/IEEE, June 1987.

[41] E. Lagnese and D. Thomas, "Architectural Partitioning for System Level Design," *Proc. 26th Design Automation Conference*, pp. 62-67, ACM/IEEE, June 1989.

[42] J. Lee, Y. Hsu, and Y. Lin, "A New Integer Linear Programming Formulation for the Scheduling Problem in Data-Path Synthesis," *ICCAD'89*, Santa Clara, CA, November 1989.

[43] M.C. McFarland, *The Value Trace: A Data Base for Automated Digital Design*, Design Research Center, Carnegie-Mellon University, Report DRC-01-4-80, December 1978.

[44] M.C. McFarland, "Using Bottom-Up Design Techniques in the Synthesis of Digital Hardware from Abstract Behavioral Descriptions," *Proceedings of the 23rd Design Automation Conference*, pp. 474-480, Las Vegas, June 1986.

[45] M.C. McFarland and T.J. Kowalski, "Incorporating Bottom-Up Design into Hardware Synthesis," *IEEE Transactions on CAD*, vol. 9, no. 9, pp. 938-950, September 1990.

[46] M.C. McFarland, A.C. Parker, and R. Camposano, "Tutorial on High-Level Synthesis," *Proceedings of the 25th Design Automation Conference*, pp. 330-336, Anaheim, California, June 1988.

[47] M.C. McFarland, A.C. Parker, and R. Camposano, "The High-Level Synthesis of Digital Systems," *Proceedings of the IEEE*, vol. 78, no. 2, pp. 301-318, February 1990.

[48] J.A. Nestor and D.E. Thomas, "Behavioral Synthesis with Interfaces," *ICCAD'86*, pp. 112-115, Santa Clara, CA, November 1986.

[49] A. Nicolau, *Percolation Scheduling: A Parallel Compilation Technique*, Ithaca, NY: Dept. of Computer Science, Cornell University, TR 85-678, May 1985.

[50] N.D. Dutt, T. Hadley, and D.D. Gajski, "An Intermediate Representation for Behavioral Synthesis," *Proceedings of the 27th Design Automation Conference*, pp. 14-19, Orlando, FL, June 1990.

[51] B.M. Pangrle and D.D. Gajski, "Design Tools for Intelligent Silicon Compilation," *IEEE Transactions on Computer-Aided Design*, vol. CAD-6, no. 6, pp. 1098-1112, November 1987.

[52] N. Park and A.C. Parker, "SEHWA: A program for Synthesis of Pipelines," *Proceedings of the 23rd Design Automation Conference*, pp. 454-460, Las Vegas, June 1986.

[53] A.C. Parker, J. Pizarro, and M. Mlinar, "MAHA: A Program for Datapath Synthesis," *Proceedings of the 23rd Design Automation Conference*, pp. 461-466, Las Vegas, June 1986.

[54] P.G. Paulin and J.P. Knight, "Force-Directed Scheduling in Automatic Data Path Synthesis," *Proceedings 24th Design Automation Conference*, pp. 195-202, Miami Beach, Florida, June 1987.

[55] P.G. Paulin and J.P. Knight, "Force-Directed Scheduling for the Behavioral Synthesis of ASIC's," *IEEE Transactions on Computer-Aided Design*, vol. 8, no. 6, pp. 661-679, June 1989.

[56] Z. Peng, "Synthesis of VLSI Systems with the CAMAD Design Aid," *Proc. 23rd Design Automation Conference*, ACM/IEEE, June 1986.

[57] R. Potasaman, J. Lis, A. Nicolau, and D. Gajski, "Percolation Based Synthesis," *Proc. 27th Design Automation Conference*, pp. 444-449, ACM/IEEE, June 1990.

[58] J. Rabaey, H. DeMan, J. Vanhoff, G. Goossens, and F. Catthoor, "Cathedral II : A Synthesis System for Multiprocessor DSP Systems," in D. Gajski, editor, *Silicon Compilation*, pp. 311-360, Addison-Wesley, 1988.

[59] W. Rosenstiel, "Optimizations in High Level Synthesis," *Microprocessing and Microprogramming (18)*, pp. 543-549, North Holland, 1986.

[60] E.A. Snow, *Automation of Module Set Independent Register-Transfer Level Design*, PhD thesis, Department of Electrical Engineering, Carnegie-Mellon University, April 1978.

[61] D.L. Springer and D.E. Thomas, "Exploiting the Special Structure of Conflict and Compatibility Graphs in High-Level Synthesis," *Proceedings ICCAD'90*, pp. 254-267, Santa Clara, CA, November 1990.

[62] D.E. Thomas, C.Y. Hitchcock III, T.J. Kowalski, J.V. Rajan, and R.A. Walker, "Methods of Automated Data Path Synthesis," *IEEE Computer*, pp. 59-70, December 1983.

[63] H. Trickey, "Flamel: A High-Level Hardware Compiler," *IEEE Transactions on Computer-Aided Design*, vol. CAD-6, no. 2, pp. 259-269, March 1987.

[64] C.J. Tseng and D.P. Siewiorek, "Automated Synthesis of Data Paths in Digital Systems," *IEEE Transactions on Computer-Aided Design*, vol. CAD-5, no. 3, pp. 379-395, July 1986.

[65] C.J. Tseng, R.S. Wei, S.G. Rothweiler, M. Tong, and A.K. Bose, "Bridge: A Versatile Behavioral Synthesis System," *25th ACM/IEEE Design Automation Conference*, pp. 415-420, Anaheim, CA, June 1988.

[66] K. Wakabayashi and T. Yoshimura, "A Resource Sharing Control Synthesis Method for Conditional Branches," *ICCAD'89*, pp. 62-65, Santa Clara, CA, November 1989.

[67] R.A. Walker and D.E. Thomas, "Design Representation and Transformation in the System Architect's Workbench," *Proceedings of the ICCAD'87*, pp. 166-169, Santa Clara, CA, November 1987.

[68] R.A. Walker and D.E. Thomas, "Behavioral Transformations for Algorithmic Level IC Design," *IEEE Transactions on CAD*, vol. 8, no. 10, pp. 1115-1128, October 1989.

[69] D.E. Thomas, E.D. Lagnese, R.A. Walker, J.A. Nestor, J.V. Rajan, and R.L. Blackburn, *Algorithmic and Register-Transfer Level Synthesis: The System Architect's Workbench*, Norwell, MA: Kluwer Academic Publishers, 1990.

Part II

Survey of High-Level Synthesis Systems

AT&T's Algorithms to Silicon Project

AT&T's Algorithms to Silicon Project uses McFarland's BUD to partition and schedule the design, and to generate a preliminary binding of operations to functional units. This result is passed to a new version of Kowalski's DAA, which uses it as a starting point for the final scheduling and data path synthesis. (The original DAA was Kowalski's thesis work at Carnegie Mellon.) Finally, other tools perform module binding, controller design, floorplanning, and layout. See also *Carnegie-Mellon's (Second) CMU-DA System* — McFarland and Kowalski used to be involved with that system.

Input

ISPS.

Internal Behavioral Representation

Carnegie Mellon's Value Trace (VT), a dataflow / controlflow graph.

Partitioning / Functional Unit Allocation / Scheduling — BUD

First, the operations are partitioned into logical and geometrical clusters, and enough functional units, registers, and multiplexors are assigned to each cluster that the cluster can be synthesized. The closeness measure considers common functionality, interconnect, and potential parallelism between operators.

Using detailed information on the size and timing of these resources, the operations in each cluster are scheduled into clock steps and phases using list scheduling, with operations on the (critical path — McFarland86a, longest path — McFarland86b) scheduled first.

The individual clusters are then placed using a min-cut placement algorithm, and wiring, delays, chip area, clock period, and the average cycle time are estimated.

Data Path Synthesis — DAA

Uses a rule-based expert system approach. First, prespecified entities such as memory arrays, constants, and I/O registers are bound. Then registers and operators are bound, using the partitioning and scheduling information generated by BUD as a starting point, and values are folded, using BUD's lifetime analysis. Finally, multiplexors, buses and interconnect are bound.

Module Binding, Controller Design, Floorplanning, and Layout

Dirkes' MOBY module binder is used to bind in technology-dependent modules. Geiger's controller designer is used to generate either microprogrammed or PLA-based controllers. Lava is used to compact the layout. See Kowalski85a for references to these tools.

Examples

Algorithms to Silicon Project

A signal processor and a floating point arithmetic example.

DAA

A PDP-8 code fragment, the MCS6502, and the IBM System/370.

References — The Algorithms to Silicon Project

McFarland90

Michael C. McFarland and Thaddeus J. Kowalski, "Incorporating Bottom-Up Design into Hardware Synthesis", *IEEE Trans. on CAD*, pages 938–950, September 1990.

Partitioning and scheduling using BUD, data path synthesis using DAA, a signal processor example, and the effect of ignoring wiring, layout, multiplexors, and registers in a cost model.

McFarland87

Michael C. McFarland, S.J., "Reevaluating the Design Space for Register-Transfer Hardware Synthesis", *Proc. of ICCAD'87*, pages 262–265, November 1987.

Shows the effect of ignoring wiring, layout, multiplexors, and registers in a cost model when evaluating a set of signal processor designs.

McFarland86a

Michael C. McFarland, S.J. and Thaddeus J. Kowalski, "Assisting DAA: The Use of Global Analysis in an Expert System", *Proc. of ICCD'86*, pages 482–485, October 1986.

Partitioning and scheduling using BUD, data path synthesis using DAA, and a signal processor example.

McFarland86b

Michael C. McFarland, S.J., "Using Bottom-Up Design Techniques in the Synthesis of Digital Hardware from Abstract Behavioral Descriptions", *Proc. of the 23rd DAC*, pages 474–480, June 1986.

Partitioning and scheduling using BUD, and a floating point arithmetic example.

Kowalski85a

T.J. Kowalski, D.G. Geiger, W.H. Wolf, and W. Fichtner, "The VLSI Design Automation Assistant: From Algorithms to Silicon", *IEEE Design and Test*, pages 33–43, August 1985.

Overview of the project, including scheduling and data path synthesis, module binding, controller allocation, floorplanning, and layout.

Kowalski85b

T.J. Kowalski, D.G. Geiger, W.H. Wolf, and W. Fichtner, "The VLSI Design Automation Assistant: A Birth in Industry", *Proc. of ISCAS'85*, pages 889–892, June 1985.

Overview of the project, including scheduling and data path synthesis, module binding, controller allocation, floorplanning, and layout.

References — DAA

Kowalski88

Thaddeus J. Kowalski, "The VLSI Design Automation Assistant: An Architecture Compiler", in *Silicon Compilation*, Daniel D. Gajski (Editor), pages 122–152, Addison-Wesley, 1988.

Knowledge acquisition, data path synthesis, the MCS6502, and the IBM System/370.

Kowalski85c

T.J. Kowalski and D.E. Thomas, "The VSLI Design Automation Assistant: What's in a Knowledge Base", *Proc. of the 22nd DAC*, pages 252–258, June 1985.

Data path synthesis phases, and a PDP-8 code fragment example.

Kowalski85d

T.J. Kowalski, *An Artificial Intelligence Approach to VLSI Design*, Kluwer Academic Publishers, 1985.

Same as Kowalski84a.

Kowalski84a

Thaddeus Julius Kowalski, *The VLSI Design Automation Assistant: A Knowledge-Based Expert System*, PhD Thesis, ECE Dept., CMU, April 1984.

Knowledge acquisition, data path synthesis, the MCS6502, and the IBM System/370.

Kowalski84b

T.J. Kowalski and D.E. Thomas, "The VSLI Design Automation Assistant: An IBM System/370 Design", *IEEE Design and Test*, pages 60–69, February 1984.

Knowledge acquisition, the MCS6502, and the IBM System/370.

Kowalski83a

T.J. Kowalski and D.E. Thomas, "The VSLI Design Automation Assistant: Prototype System", *Proc. of the 20th DAC*, pages 479–483, June 1983.

Overview and the MCS6502.

Kowalski83b

T.J. Kowalski and D.E. Thomas, "The VSLI Design Automation Assistant: Learning to Walk", *Proc. of ISCAS'83*, May 1983.

Knowledge acquisition and the MCS6502.

Kowalski83c

T.J. Kowalski and D.E. Thomas, "The VSLI Design Automation Assistant: First Steps", *Proc. of the 26th Compcon*, pages 126–130, February 1983.

Knowledge acquisition, the DAA prototype, and the MCS6502.

AT&T's Bridge System

AT&T's Bridge system uses Tseng's Facet methodology to allocate the data path. Bridge includes scheduling, data path synthesis, module binding, and controller design. See also *Carnegie Mellon's Facet / Emerald System* (Tseng used to be involved with that system), and *AT&T's CHARM System*.

Input

FDL2.

Internal Behavioral Representation

A dataflow / controlflow graph.

Scheduling

Uses ASAP scheduling. Supports local slicing (a control state appears in only one basic block) as well as global slicing (a control state spans basic blocks — either across CASE branches or concurrent branches). Global slicing also facilitates the synthesis of concurrent processes, although the data path synthesis algorithms described below have to be modified somewhat.

Data Path Synthesis

Register Binding

Builds a compatibility graph, with one node per variable, and an edge connecting variables that can be assigned to the same register (variables that are live in different control steps). Then finds the minimum number of cliques, thus minimizing the number of registers.

Uses as many ALUs as necessary.

Functional Unit Binding

Builds a compatibility graph, with one node per data operator, and an edge connecting operators that can be assigned to the same ALU (operators not used in the same control step). Then finds the minimum number of cliques, thus minimizing the number of ALUs. As the cliques are found, the algorithm gives priority to grouping together operators with (1) the same inputs and outputs, and (2) performing the same operation.

Interconnection

Builds a compatibility graph, with one node per interconnection variable, and an edge connecting variables that can never be used simultaneously. Then finds the minimum number of cliques, thus minimizing the the number of interconnection units.

Module Binding

Tseng89 — Mind

Generates gate-level logic structures for the data path (single and multiple-function ALUs, registers, and multiplexors) and controller.

Tseng88a, Tseng88b, Dussault84 — FDS

Uses AT&T's FDS system [Dussault84] for module binding, decomposing the design into smaller components, until each matches a standard cell implementation in the data base. ALUs can either be designed by refining existing ALUs, or designed from scratch based on Boolean equations or a truth table.

Controller Design

Uses AT&T's FSM Synthesizer [Tseng86a] to design the controller. Don't care conditions are used to allow column compaction, reducing the number of controller outputs.

Examples

The Intel 8251.

References

Tseng89

Chia-Jeng Tseng, Steven G. Rothweiler, Shailesh Sutarwala, and Ajit M. Prabhu, “Mind: A Module Binder for High Level Synthesis”, *Proc. of ICCD'89*, pages 420–423, October 1989.

Module binding.

Tseng88a

Chia-Jeng Tseng, Ruey-Sing Wei, Steven G. Rothweiler, Michael M. Tong, and Ajoy K. Bose, “Bridge: A Versatile Behavioral Synthesis System”, *Proc. of the 25th DAC*, pages 415–420, June 1988.

System overview, local and global slicing, and the Intel 8251.

Tseng88b

Chia-Jeng Tseng, Ruey-Sing Wei, Steven G. Rothweiler, Michael M. Tong, and Ajoy K. Bose, “Bridge: A Behavioral Synthesis System for VLSI”, *Proc. of the 1988 Custom Integrated Circuits Conf.*, pages 2.6.1–2.6.4, May 1988.

System overview.

Wei87

Ruey-Sing Wei and Chia-Jeng Tseng, “Column Compaction and Its Application to the Control Path Synthesis”, *Proc. of ICCAD'87*, pages 320–323, November 1987.

Two methods for column compaction.

Tseng86a

Chia-Jeng Tseng, Ajit M. Prabhu, Cedric Li, Zafer Mehmood, and Michael M. Tong, “A Versatile Finite State Machine Synthesizer”, *Proc. of ICCAD'86*, pages 206–209, November 1986.

Overview of the FSM Synthesizer, a system for synthesizing finite state machines using either polycells, PALs, or PLAs.

Dussault84

Jean Dussault, Chi-Chang Liaw, and Michael M. Tong, "A High Level Synthesis Tool for MOS Chip Design", *Proc. of the 21st DAC*, pages 308–314, June 1984.

Overview of FDS, a system for synthesizing MSI cells using other cells and FDS primitives.

AT&T's CHARM System

AT&T's CHARM (Concurrent Hardware Allocation by Repeated Merging) system uses AT&T's Bridge system for scheduling. CHARM (previously called SAM) then performs the data path synthesis. See also *AT&T's Bridge System*.

Input

FDL2. See *AT&T's Bridge system*.

Internal Behavioral Representation

A dataflow / controlflow graph. See *AT&T's Bridge system*.

Scheduling

Woo90

Uses AT&T's Bridge system.

Shin89

First, the algorithm starts with an initial control step schedule, for a fixed number of control steps. Then the cost is calculated, based on the cost of the functional units and their usage, the gradient direction of the cost is determined, and the schedule is modified to lower the cost. This process continues until the algorithm converges.

Data Path Synthesis

Registers, functional units and interconnect (wires, multiplexors) are allocated simultaneously by an iterative / constructive algorithm that considers the cost of registers, single-function units, multiple-function units, and interconnect. Initially, every operation is in a different disjoint operator set. During each iteration, the algorithm considers the savings that would result from merging each pair of disjoint operator sets. For the pair with the maximum possible savings, the operator sets are merged, and the data path is updated.

Examples

A differential equation example, three code segment examples, MAHA's example, and a fifth-order digital elliptic wave filter example.

References

Woo90

Nam-Sung Woo, "A Global, Dynamic Register Allocation and Binding for a Data Path Synthesis System", *Proc. of the 27th DAC*, pages 505–510, June 1990.

Data path synthesis, including register allocation and binding, and three code segment examples.

Shin89

Hyunchul Shin and Nam S. Woo, "A Cost Function Based Optimization Technique for Scheduling in Data Path Synthesis", *Proc. of ICCD'89*, pages 424–427, October 1989.

Control step scheduling, MAHA's example, and a fifth-order digital elliptic wave filter example.

Woo89

Nam-Sung Woo and Hyunchul Shin, "A Technology-Adaptive Allocation of Functional Units and Connections", *Proc. of the 26th DAC*, pages 602–605, June 1989.

Functional unit and interconnect allocation, and a differential equation example.

Carleton's Elf System

The Elf system was Girczyc's thesis work at Carleton University; it was further refined at Audesyn Inc. Elf includes scheduling and data path synthesis.

Input

A subset of Ada.

Internal Behavioral Representation

A dataflow / controlflow graph (CDFG).

Scheduling / Data Path Synthesis

List scheduling, with operations ranked according to their urgency (the minimum number of cycles required between the operation and any enclosing timing constraints).

As operations are placed into each control step, graph grammar productions are used to allocate functional units and registers. If more than one production matches an operator, the one resulting in the lowest cost (including multiplexors and interconnect) is applied.

As interconnect is considered for each control step, the most constrained unallocated data transfer is allocated using the minimum cost path, then the next most constrained unallocated data transfer is chosen, etc.

After scheduling and allocation, a greedy global partitioning / clustering algorithm is used to optimize the interconnect.

Examples

A temperature controller example, and a fifth-order digital elliptic wave filter example.

References

Ly90

Tai A. Ly, W. Lloyd Elwood, and Emil F. Girczyc, "A Generalized Interconnect Model for Data Path Synthesis", *Proc. of the 27th DAC*, pages 168–173, June 1990.

Interconnect allocation, and a fifth-order digital elliptic wave filter example.

Girczyc87

E.F. Girczyc, "Loop Winding — A Data Flow Approach to Functional Pipelining", *Proc. of ISCAS'87*, pages 382–385, May 1987.

Processing a dataflow graph to pipeline the design.

Girczyc85

Emil F. Girczyc, Ray J.A. Buhr, and John P. Knight, "Applicability of a Subset of Ada as an Algorithmic Hardware Description Language for Graph-Based Hardware Compilation", *IEEE Trans. on CAD*, pages 134–142, April 1985.

The Ada subset, and a temperature controller example.

Girczyc84

E .F. Girczyc and J.P. Knight, "An Ada to Standard Cell Hardware Compiler Based on Graph Grammars and Scheduling", *Proc. of ICCD'84*, pages 726–731, October 1984.

Scheduling, data path synthesis, and a temperature controller example.

Carleton's HAL System

The HAL (Hardware Allocator) system was Paulin's thesis work at Carleton University. HAL includes scheduling, data path synthesis, and design iteration.

Input

A manually entered dataflow / controlflow graph.

Internal Behavioral Representation

Uses a dataflow / controlflow graph.

Scheduling

Paulin89[a,b,c]

Combines the force-directed scheduling (FDS) algorithm presented in Paulin88 with list scheduling to produce a force-directed list scheduling (FDLS) algorithm. Whereas force-directed scheduling constrains the length of the schedule, list-scheduling (including force-directed list scheduling) constrains the number of functional units. In the FDLS algorithm, force is used as the priority function — of the set of ready operations, the operation with the lowest force is deferred (not considered for the current control step), and this process is continued until the constraint on the number of functional units is met.

The FDS and FDLS algorithms can also be used together to explore the design space. The FDS algorithm is used first with a maximum time constraint to find a near-optimal allocation; then the FDLS algorithm is used with that allocation to try to find a faster schedule.

Paulin88

Done by the InHAL module, using a force-directed scheduling algorithm that tries to schedule the operations so as to balance the distribution of operations using the same hardware resources, without lengthening the schedule. First, both an ASAP schedule and an ALAP schedule are generated, and the results combined to indicate the possible control steps for each operation. Then, assuming an equal 1/nth probability of each operation being scheduled into one of the n possible control steps, for each type of operation for each control step the probabilities are added to produce a set of distribution graphs. These distribution graphs are then used to calculated a force for each operation for each possible control step; this force is positive if that schedule causes an increase in operation concurrency, and negative if it causes a decrease. Extra forces are added to consider the effect of the schedule on all other operations. Once all forces have been calculated, the operation / time step with the most negative (or lowest positive) force is scheduled, the distribution graphs and forces are updated, and the process continues.

Constraints can be placed on the total hardware cost, or on the number of functional units of each type.

Paulin86

Done by the InHAL module, first scheduling the operations on the critical path, then the others. For the others, both an ASAP schedule and an ALAP schedule are generated, and the results combined to indicate the possible control steps for each operation. Operations are distributed when possible to minimize the scheduling of similar operations into the same time step. The MidHAL module is then called to estimate the allocation of operations to functional units, and operations are reassigned when necessary to minimize the scheduling of operations using similar resources into the same time step. *Note that there is no mention here of force-directed scheduling.*

Data Path Synthesis

The MidHAL module allocates a set of functional units to perform the scheduled operations, but does not bind specific operations to specific functional units. This allocation is done by a rule-based expert system, which takes into account the available cells, their area cost, and the timing constraints.

The ExHAL module then binds the operations to specific functional units and completes the data path. First, the operations are bound to specific functional units using a greedy algorithm, attempting to minimize the interconnect by minimizing the number of different sources and destinations bound to a functional unit. Second, storage operations are inserted between data-dependent operations in different control steps, registers are assigned to the storage operations, and disjoint registers are merged using a clique partitioning algorithm. Third, multiplexors and interconnect are added where necessary, and the multiplexors are also merged using clique partitioning. Finally, local optimizations are applied to improve the resulting design.

Design Iteration

Phase 1 — Default allocation: single-function functional units are allocated to perform each type of operation.

Phase 2 — Preliminary schedule: the InHAL module balances the distribution of operations that use the same hardware resources, using the default allocation.

Phase 3 — Refined allocation: the MidHAL module performs allocation of single- and multi-function functional units, sometimes substituting slower functional units or combining multiple operations into a multi-function functional unit.

Phase 4 — Final schedule: the InHAL module balances the distribution of operations using similar resources, using the refined allocation.

Examples

MAHA's example (originally from Park), a pipelined 16-point digital FIR filter example, a temperature controller example, a fifth-order digital elliptic wave filter example, a second-order differential equation example, and Tseng's running example.

References

Paulin89a

Pierre G. Paulin and John P. Knight, "Algorithms for High-Level Synthesis", *IEEE Design and Test*, pages 18–31, December 1989.

Force-directed list scheduling, a second-order differential equation example, and a fifth-order digital elliptic wave filter example.

Paulin89b

Pierre G. Paulin and John P. Knight, "Scheduling and Binding Algorithms for High-Level Synthesis", *Proc. of the 26th DAC*, pages 1–6, June 1989.

Force-directed list scheduling, a second-order differential equation example, and a fifth-order digital elliptic wave filter example.

Paulin89c

Pierre G. Paulin and John P. Knight, "Force-Directed Scheduling for the Behavioral Synthesis of ASIC's", *IEEE Trans. on CAD*, pages 661–679, June 1989.

Other control step schedulers, force-directed scheduling, force-directed list scheduling, a second-order differential equation example, a pipelined 16-point digital FIR filter example, and a fifth-order digital elliptic wave filter example.

Paulin88

Pierre G. Paulin, *High-Level Synthesis of Digital Circuits Using Global Scheduling and Binding Algorithms*, PhD Thesis, Dept. of Electronics, Carleton University, January 1988.

Other synthesis systems, scheduling and data path synthesis, design iteration, a second-order differential equation example, MAHA's example (originally from Park), a pipelined 16-point digital FIR filter example, Tseng's running example, and a fifth-order digital elliptic wave filter example.

Paulin87

P.G. Paulin and J.P. Knight, "Force-Directed Scheduling in Automatic Data Path Synthesis", *Proc. of the 24th DAC*, pages 195–202, June 1987.

Force-directed scheduling, MAHA's example (originally from Park), a pipelined 16-point digital FIR filter example, a temperature controller example, and a fifth-order digital elliptic wave filter example.

Paulin86

P.G. Paulin, J.P. Knight, and E.F. Girczyc, "HAL: A Multi-Paradigm Approach to Automatic Data Path Synthesis", *Proc. of the 23rd DAC*, pages 263–270, June 1986.

Scheduling and data path synthesis, a second-order differential equation example, and Tseng's running example.

Carnegie Mellon's (First) CMU-DA System

This was the first of two Carnegie Mellon University Design Automation (CMU-DA) systems. This CMU-DA system included scheduling, data path synthesis, controller design, module binding, and linking between the algorithmic level input description and the Register Transfer level design. See also *Carnegie-Mellon's (Second) CMU-DA System* and *Carnegie-Mellon's System Architect's Workbench* — both are successors to this system.

Input

ISPS.

Scheduling, Data Path Synthesis, Controller Design, and Module Binding

Not described here.

Examples

The Mark1, the AM2909, the AM2910, the PDP-11/40, and the PDP-8.

References

Thomas83

Donald E. Thomas and John A. Nestor, "Defining and Implementing a Multilevel Design Representation With Simulation Applications", *IEEE Trans. on CAD*, pages 135–145, July 1983.

Linking the ISPS behavioral descriptions, the microsequence table, and the structural representation, and applications in timing abstraction.

Nestor82

J.A. Nestor and D.E. Thomas, "Defining and Implementing a Multilevel Design Representation with Simulation Applications", *Proc. of the 19th DAC*, pages 740–746, June 1982.

Linking the ISPS behavioral descriptions, the microsequence table, and the structural representation, and applications in timing abstraction.

Hafer81a

Louis J. Hafer, *Automated Data-Memory Synthesis: A Formal Method for the Specification, Analysis, and Design of Register-Transfer Level Digital Logic*, PhD Thesis, Dept. of Electrical Engineering, Carnegie Mellon University, December 1981.

A formal Register-Transfer level hardware model, and automatic data path synthesis using linear programming techniques.

DiRusso81

Renato Di Russo, *A Design Implementation Using the CMU-DA System*, Master's Thesis, Dept. of Electrical Engineering, Carnegie Mellon University, October 1981.

Using the CMU-DA system to design a PDP-8, and errors detected in the system.

Thomas81a

Donald E. Thomas, "The Automatic Synthesis of Digital Systems", *Proc. of the IEEE*, pages 1200–1211, October 1981.

Nestor81

John A. Nestor, *Defining and Implementing a Multilevel Design Representation With Simulation Applications*, Master's Thesis, Dept. of Electrical Engineering, Carnegie Mellon University, September 1981.

Linking the ISPS behavioral descriptions, the microsequence table, the structural representation, and applications in timing abstraction.

Hertz81

Lois Hertz, *Automatic Synthesis of Control Sequencers*, Master's Thesis, Dept. of Electrical Engineering, Carnegie Mellon University, September 1981.

Generation of PLA-based controllers.

Director81

Stephen W. Director, Alice C. Parker, Daniel P. Siewiorek, and Donald E. Thomas, Jr., "A Design Methodology and Computer Aids for Digital VLSI Systems", *IEEE Trans. on Circuits and Systems*, pages 634–635, July 1981.

Overview of the first and second CMU-DA system (mixed together), including placement and routing.

Leive81a

G.W. Leive and D.E. Thomas, "A Technology Relative Logic Synthesis and Module Selection System", *Proc. of the 18th DAC*, pages 479–485, June 1981.

Automatic module binding and transformation to meet constraints, design space exploration, and a PDP-8 example.

Nagle81

Andrew W. Nagle and Alice C. Parker, "Algorithms for Multiple-Criterion Design of Microprogrammed Control Hardware", *Proc. of the 18th DAC*, pages 486–493, June 1981.

Microword encoding and the PDP-11/40.

Hafer81b

Louis Hafer and Alice C. Parker, "A Formal Method for the Specification, Analysis, and Design of Register-Transfer Level Digital Logic", *Proc. of the 18th DAC*, pages 846–853, June 1981.

A formal Register-Transfer level hardware model, and automatic data path synthesis using linear programming techniques.

Leive81b

Gary W. Leive, *The Design, Implementation, and Analysis of an Automated Logic Synthesis and Module Selection System*, PhD Thesis, Dept. of Electrical Engineering, Carnegie Mellon University, January 1981.

Register-Transfer level transformations, automatic module binding and transformation to meet constraints, design space exploration, and a PDP-8 example.

Thomas81b

Donald E. Thomas and Daniel P. Siewiorek, "Measuring Design Performance to Verify Design Automation Systems", *IEEE Trans. on Computers*, January 1981.

Nagle80

Andrew W. Nagle, *Automated Design of Digital-System Control Sequencers from Register-Transfer Specifications*, PhD Thesis, Dept. of Electrical Engineering, Carnegie Mellon University, December 1980.

Control graph and module control notation, scheduling and microword encoding, and the PDP-11/40.

Cloutier80

Richard J. Cloutier, *Control Allocation: The Automated Design of Digital Controllers*, Master's Thesis, Dept. of Electrical Engineering, Carnegie Mellon University, April 1980.

Control graph generation, optimization, scheduling, the Mark1, the AM2909, the AM2910, and the PDP-11/40.

Parker79

A. Parker, D. Thomas, D. Siewiorek, M. Barbacci, L. Hafer, G. Leive, and J. Kim, "The CMU Design Automation System: An Example of Automated Data Path Design", *Proc. of the 16th DAC*, pages 73–80, June 1979.

Automatic synthesis of the PDP-8/E.

Nagle78

Andrew W. Nagle, "Automatic Design of Micro-Controllers", *Proc. 11th Microprogramming Conf.*, 1978.

Hafer78

Louis Hafer and Alice C. Parker, "Register Transfer Level Automatic Digital Design: The Allocation Process", *Proc. of the 15th DAC*, June 1978.

Hafer77

Louis Hafer, *Data-Memory Allocation in the Distributed Logic Design Style*, Master's Thesis, Dept. of Electrical Engineering, Carnegie Mellon University, December 1977.

Thomas77a

D.E. Thomas and D.P. Siewiorek, "A Technology-Relative Computer-Aided Design System: Abstract Representations, Transformations and Design Tradeoffs", *Proc. of the 14th DAC*, June 1977.

Thomas77b

Donald E. Thomas, *The Design and Analysis of an Automated Design Style Selector*, PhD Thesis, Dept. of Electrical Engineering, Carnegie Mellon University, March 1977.

Carnegie Mellon's (Second) CMU-DA System

This was the second of two Carnegie Mellon University Design Automation (CMU-DA) systems. This CMU-DA system included scheduling, data path synthesis, controller design, module binding, and linking between the algorithmic level and Register Transfer level behavioral representations. See also *Carnegie-Mellon's (First) CMU-DA System* and *Carnegie-Mellon's System Architect's Workbench* — the predecessor and successor to this system. See also *AT&T's Algorithms to Silicon Project* — McFarland and Kowalski were later involved with that project.

Input

ISPS.

Internal Behavioral Representation

Uses the Value Trace (VT), a dataflow / controlflow graph.

Transformations

Supports behavioral transformations, primarily to improve the efficiency of the control structure: constant folding, common subexpression elimination, dead procedure elimination, inline expansion and formation of procedures, code motion into and out of the branches of decoding operations, and loop unrolling.

Scheduling

ASAP scheduling on a basic-block basis.

Data Path Synthesis

EMUCS

The EMUCS data path allocator uses an algorithm originally defined by McFarland, but unpublished. It attempts to bind dataflow elements onto hardware elements in a step-by-step manner, considering all unbound dataflow elements in each step. To decide which element to bind, EMUCS maintains a cost table, listing the cost of binding each unbound dataflow element onto each hardware element. For each unbound dataflow element, EMUCS calculates the difference of the two lowest binding costs, then binds the dataflow element with the highest difference to the hardware element with the lowest cost. This attempts to minimize the additional cost that would be incurred if that element were bound in a later step.

DAA

The DAA (Design Automation Assistant) uses a knowledge-based expert system (KBES) approach to data path synthesis. First, it allocates memories, global registers, formal parameters, and constants. Second, it partitions the design, and allocates clock phases, operators, registers, data paths, and control logic within each block. Third, local optimizations are applied to remove or combine registers, combine modules, etc. Finally, global optimizations are applied to remove unreferenced modules, unneeded registers, reduce multiplexor trees, and allocate bus structures.

Examples

The Mark1, the MCS6502, and the IBM System/370.

References — The Overall System

Thomas83

D.E. Thomas, C.Y. Hitchcock III, T.J. Kowalski, J.V. Rajan, and R.A. Walker, "Methods of Automatic Data Path Synthesis", *IEEE Computer*, pages 59–70, December 1983.

System overview, transformations, scheduling, data path synthesis using EMUCS and DAA, multilevel representation, and the MCS6502.

Director81

Stephen W. Director, Alice C. Parker, Daniel P. Siewiorek, and Donald E. Thomas, Jr., "A Design Methodology and Computer

Aids for Digital VLSI Systems", *IEEE Trans. on Circuits and Systems*, pages 634–635, July 1981.

Overview of the first and second CMU-DA system (mixed together), including placement and routing.

References — DAA

Kowalski88

Thaddeus J. Kowalski, "The VLSI Design Automation Assistant: An Architecture Compiler", in *Silicon Compilation*, Daniel D. Gajski (Editor), pages 122–152, Addison-Wesley, 1988.

Knowledge acquisition, data path synthesis, the MCS6502, and the IBM System/370.

Kowalski85a

T.J. Kowalski and D.E. Thomas, "The VSLI Design Automation Assistant: What's in a Knowledge Base", *Proc. of the 22nd DAC*, pages 252–258, June 1985.

Data path synthesis phases, and a PDP-8 code fragment example.

Kowalski85b

T.J. Kowalski, *An Artificial Intelligence Approach to VLSI Design*, Kluwer Academic Publishers, 1985.

Same as Kowalski84a.

Kowalski84a

Thaddeus Julius Kowalski, *The VLSI Design Automation Assistant: A Knowledge-Based Expert System*, PhD Thesis, ECE Dept., CMU, April 1984.

Knowledge acquisition, data path synthesis, the MCS6502, and the IBM System/370.

Kowalski84b

T.J. Kowalski and D.E. Thomas, "The VSLI Design Automation Assistant: An IBM System/370 Design", *IEEE Design and Test*, pages 60–69, February 1984.

Knowledge acquisition, the MCS6502, and the IBM System/370.

Kowalski83a

T.J. Kowalski and D.E. Thomas, "The VSLI Design Automation Assistant: Prototype System", *Proc. of the 20th DAC*, pages 479–483, June 1983.

Overview and the MCS6502.

Kowalski83b

T.J. Kowalski and D.E. Thomas, "The VSLI Design Automation Assistant: Learning to Walk", *Proc. of ISCAS'83*, pages 186–190, May 1983.

Knowledge acquisition and the MCS6502.

Kowalski83c

T.J. Kowalski and D.E. Thomas, "The VSLI Design Automation Assistant: First Steps", *Proc. of the 26th Compcon*, pages 126–130, February 1983.

Knowledge acquisition, the DAA prototype, and the MCS6502.

References — Master's Projects

Thomas87

Donald E. Thomas, Robert L. Blackburn, and Jayanth V. Rajan, "Linking the Behavioral and Structural Domains of Representation", *IEEE Trans. on CAD*, pages 103–110, January 1987.

Linking values in the GDB (ISPS parse tree) and the Value Trace, and linking the behavioral and structural domains.

Blackburn85

Robert L. Blackburn and Donald E. Thomas, "Linking the Behavioral and Structural Domains of Representation in a Synthesis System", *Proc. of the 22nd DAC*, pages 374–380, June 1985.

Linking values in the GDB (ISPS parse tree) and the Value Trace.

Dirkes85

Elizabeth Dirkes, *A Module Binder for the CMU-DA System,* Master's Thesis, Dept. of Electrical and Computer Engineering, Carnegie Mellon University, May 1985.

Module binding, the Mark1, the AM2910, a sequencer example, and the MCS6502.

Geiger84

David John Geiger, *A Framework for the Automatic Design of Controllers*, Dept. of Electrical and Computer Engineering, Carnegie Mellon University, July 1984.

Design of microcoded and PLA-based controllers, a multiplier example, the Mark1, the MCS6502, a sequencer example, and the RCA1802.

Blackburn84

Robert L. Blackburn, *Linking Behavioral Representations in an IC Design System*, Dept. of Electrical and Computer Engineering, Carnegie Mellon University, May 1984.

Linking values in the GDB (ISPS parse tree) and the Value Trace.

Hitchcock83a

Charles Y. Hitchcock III and Donald E. Thomas, "A Method of Automatic Data Path Synthesis", *Proc. of the 20th DAC*, pages 484–489, June 1983.

Data path synthesis, the PDP-8, and the MCS6502.

Hitchcock83b

Charles Y. Hitchcock III, *A Method of Automatic Data Path Synthesis*, Master's Thesis, Dept. of Electrical and Computer Engineering, Carnegie Mellon University, January 1983.

Data path synthesis, the Mark1, the PDP-8, and the MCS6502.

Walker83

Robert A. Walker and Donald E. Thomas, "Behavioral Level Transformation in the CMU-DA System", *Proc. of the 20th DAC*, pages 788–789, June 1983.

The implementation of some of Snow's transformations [Snow78], the definition and implementation of some new transformations, and the Intel 8080.

Walker82

Robert A. Walker, *A Transformation Package for the Behavioral Level of the CMU-DA System*, Master's Thesis, Dept. of Electrical Engineering, Carnegie Mellon University, October 1982.

The implementation of some of Snow's transformations [Snow78], the definition and implementation of some new transformations, the AM2901, the AM2903, and the Intel 8080.

Vasantharajan82

Jayanth Vasantharajan, *Design and Implementation of a VT-Based Multi-Level Representation*, Master's Thesis, Dept. of Electrical Engineering, Carnegie Mellon University, February 1982.

A multilevel representation to describe structure and control specification.

Lertola81

John F. Lertola, *Design and Implementation of a Hardware Synthesis Design Aid*, Master's Thesis, Dept. of Electrical Engineering, Carnegie Mellon University, October 1981.

Interactive data path synthesis.

Gatenby81

David A. Gatenby, *Digital Design from an Abstract Algorithmic Representation: Design and Implementation of a Framework for Interactive Design*, Master's Thesis, Dept. of Electrical Engineering, Carnegie Mellon University, October 1981.

Organization of the Value Trace based synthesis system, parsing a Value Trace file to build a set of data structures, ASAP control step scheduling, a graphical Value Trace display, and Value Trace metrics.

References — The Value Trace

McFarland81a

Michael C. McFarland, S.J., *Mathematical Models for Formal Verification in a Design Automation System*, PhD Thesis, Dept. of Electrical Engineering, Carnegie Mellon University, July 1981.

A formal behavioral model for subsets of ISPS and the Value Trace, and proof of correctness for some of Snow's transformations.

McFarland81b

Michael C. McFarland, S.J., On Proving the Correctness of Optimizing Transformations in a Digital Design Automation System, *Proc. of the 18th DAC*, pages 90–97, June 1981.

A formal behavioral model for subsets of ISPS and the Value Trace, and proof of correctness for inline expansion.

McFarland78

Michael C. McFarland, S.J., *The Value Trace: A Data Base for Automated Digital Design*, Master's Thesis, Dept. of Electrical Engineering, Carnegie Mellon University, December 1978.

Parsing an ISPS description to produce a Value Trace, and some global transformations on the Value Trace.

Snow78a

E. Snow, D. Siewiorek, and D. Thomas, A Technology-Relative Computer Aided Design System: Abstract Representations, Transformations, and Design Tradeoffs, *Proc. of the 15th DAC*, June 1978.

Snow78

Edward A. Snow, *Automation of Module Set Independent Register-Transfer Level Design*, PhD Thesis, Dept. of Electrical Engineering, Carnegie Mellon University, April 1978.

The Value Trace, transformations on the Value Trace, and design space exploration using the transformations (on paper) to produce various PDP-11 architectures.

Carnegie Mellon's System Architect's Workbench

Carnegie Mellon University's System Architect's Workbench is the successor to the first and second Carnegie Mellon University Design Automation (CMU-DA) systems. The Workbench supports three synthesis paths: a general synthesis path, using the transformations, APARTY, CSTEP, and EMUCS; a pipelined-instruction-set-processor-specific synthesis path, using SAM; and a microprocessor-specific synthesis path, using SUGAR. See also *Carnegie Mellon's (First) CMU-DA System* and *Carnegie Mellon's (Second) CMU-DA System*, the predecessors of this system.

Input

ISPS, extended to support processes and message-passing inter-process communication, and user-definable operations.

Verilog.

VHDL, using CAD Language Systems, Inc.'s parser.

Transformations / APARTY / CSTEP / EMUCS Synthesis Path

Internal Behavioral Representation

Uses the Value Trace (VT), a dataflow / controlflow graph.

Transformations

Supports behavioral and structural transformations. Some transformations are used primarily to improve the efficiency of the control structure: inline expansion and formation of procedures, code motion into and out of the branches of decoding operations, combination of nested decoding operations, etc. Other transformations allow the designer to explore Algorithmic level design alternatives: adding concurrent processes to a design, pipelining a design, and structurally partitioning a design.

Partitioning — APARTY

The APARTY architectural partitioner supports instruction set partitioning and intra-chip partitioning (to obtain additional operator-level parallelism between the partitions). It uses a multistage clustering algorithm, with the stages applied sequentially and each stage using the results from the previous stage. For an instruction set architecture, control clustering groups operators into instructions, and data flow clustering groups the instructions that use common data carriers together. Inter-procedure clustering groups instructions groups with their auxiliary procedures, and common procedure clustering groups instruction groups that use common auxiliary procedures.

Scheduling — CSTEP

The CSTEP control step scheduler uses list scheduling on a block-by-block basis, with timing constraint evaluation as the priority function. Operations are scheduled into control steps one basic block at a time, with the blocks scheduled in execution order using a depth-first traversal of the control flow graph. For each basic block, data ready operator are considered for placement into the current control step, using a priority function that reflects whether or not that placement will violate timing constraints. Resource limits may be applied to limit the number of operators of a particular type in any one control step.

Data Path Synthesis — EMUCS

The EMUCS data path allocator uses an algorithm originally defined by McFarland, but unpublished. It attempts to bind dataflow elements onto hardware elements in a step-by-step manner, considering all unbound dataflow elements in each step. To decide which element to bind, EMUCS maintains a cost table, listing the cost of binding each unbound dataflow element onto each hardware element. For each unbound dataflow element, EMUCS calculates the difference of the two lowest binding costs, then binds the dataflow element with the highest difference to the hardware element with the lowest cost. This attempts to minimize the additional cost that would be incurred if that element were bound in a later step.

In a post-processing phase, Busser then adds buses to the design, replacing multiplexors as necessary. A compatibility graph is constructed, representing all inputs to each module in the data path, with an edge between the nodes if the inputs can not share a bus. Tseng's clique partitioning algorithm is then used to produce buses, merging nodes with a maximum number of common neighbors.

SAM Synthesis Path

Internal Behavioral Representation

Uses the Value Trace (VT), a dataflow / controlflow graph.

Scheduling / Data Path Synthesis

SAM (Scheduling, Allocation, and Mapping) uses concepts taken from force-directed scheduling to perform scheduling and data path synthesis. First, ASAP and ALAP scheduling is used to determine the possible control steps for each operation. A force is then calculated for the scheduling of each operation into each possible control step; this force is based on the operation, the control step, and the available functional units. The operation / control step pair with the most-negatively-value force is then chosen, and bound to the functional unit with the lowest instance cost and connection compatibility.

SAM does not yet support register allocation and binding.

SUGAR Synthesis Path

Internal Behavioral Representation

Uses a tree representation and a dataflow representation.

Scheduling / Data Path Synthesis

The SUGAR system uses both algorithms and rules to encode knowledge about microprocessor design. SUGAR divides the microprocessor synthesis problem into phases, each operating on a common tree representation. These phases improve the efficiency of the control step schedule, select a bus structure, allocate registers and functional units, select microcode to implement the behavior, and improve the design using cost / speed tradeoffs.

Examples

Transformations / APARTY / CSTEP / EMUCS Synthesis Path

The Intel 8251, a Multibus example, a protocol adapter example, a block transfer example, a fifth-order digital elliptic wave filter example, a Kalman filter example, the BTL310, an ADPCM, the Risc-1, the MCS6502, and the IBM System/370.

SAM Synthesis Path

A fifth-order digital elliptic wave filter example.

SUGAR Synthesis Path

The MCS6502, and the MC68000.

References

Springer90

D.L. Springer and D.E. Thomas, "Exploiting the Special Structure of Conflict and Compatibility Graphs in High-Level Synthesis", *Proc. of ICCAD'90*, pages 254–257, November 1990.

Exploiting conflict and compatibility graphs, and bus-merging using a compatibility graph.

Cloutier90

Richard J. Cloutier and Donald E. Thomas, "The Combination of Scheduling, Allocation, and Mapping in a Single Algorithm", *Proc. of the 27th DAC*, pages 71–76, June 1990.

Scheduling and data path synthesis, and a fifth-order digital elliptic wave filter example.

Thomas90

D.E. Thomas, E.D. Lagnese, R.A. Walker, J.A. Nestor, J.V. Rajan, and R.L. Blackburn, *Algorithmic and Register-Transfer Level Synthesis: The System Architect's Workbench*, Kluwer Academic Publishers, Boston, 1990.

System overview, design representations, transformations, architectural partitioning, scheduling, data path synthesis, microprocessor synthesis, a fifth-order digital elliptic wave filter example, a Kalman filter example, the BTL310, the MCS6502, and the MC68000.

Walker89

Robert A. Walker and Donald E. Thomas, "Behavioral Transformation for Algorithmic Level IC Design", *IEEE Trans. on CAD*, pages 1115–1128, October 1989.

Algorithmic level behavioral and structural transformations, adding concurrent processes to an elliptical filter, pipelining the MCS6502.

Lagnese89

E. Dirkes Lagnese and D.E. Thomas, "Architectural Partitioning for System Level Design", *Proc. of the 26th DAC*, pages 62–67, June 1989.

Architectural partitioning, guiding CSTEP, EMUCS and Busser, and a fifth-order digital elliptic wave filter example.

Lagnese89

Elizabeth Dirkes Lagnese, *Architectural Partitioning for System Level Design of Integrated Circuits*, PhD Thesis, Dept. of Electrical and Computer Engineering, Carnegie Mellon University, March 1989.

Architectural partitioning, guiding CSTEP, EMUCS and Busser, a fifth-order digital elliptic wave filter example, a Kalman filter example, the BTL310, an ADPCM, the MCS6502, and the IBM System/370.

Rajan89

Jayanth V. Rajan, *Automatic Synthesis of Microprocessors*, PhD Thesis, Dept. of Electrical and Computer Engineering, Carnegie Mellon University, January 1989.

Register Transfer level synthesis, behavioral representation and transformation, code generation and selection, register and bus assignment, the MCS6502, and the MC68000.

Murgai89

Rajeev Murgai, *Automatic Design of Control Paths for System Level Synthesis*, Master's Thesis, Dept. of Electrical and Computer Engineering, Carnegie Mellon University, January 1989.

Automatic design of PLA-based, random logic, and microprogrammed controllers.

Un89

Edware D. Un, *Evaluating Register-Transfer Level Synthesis Tools Using Gate Level Simulation*, Master's Thesis, Dept. of Electrical and Computer Engineering, Carnegie Mellon University, January 1989.

Module binding, and interfacing to Seattle Silicon Corporation's ChipCrafter placement and routing tool.

Blackburn88

Robert L. Blackburn, *Relating Design Representations in an Automated IC Design System*, PhD Thesis, Dept. of Electrical and Computer Engineering, Carnegie Mellon University, October 1988.

Other linking systems, linking values and operators in the GDB (ISPS parse tree), the Value Trace, the allocated hardware, and the control step schedule.

Thomas88

D.E. Thomas, E.M. Dirkes, R.A. Walker, J.V. Rajan, J.A. Nestor, and R.L. Blackburn, "The System Architect's Workbench", *Proc. of the 25th DAC*, pages 337–343, June 1988.

Overview of the system, transformations, control step scheduling, architectural partitioning, synthesis using

SUGAR, a fifth-order digital elliptic wave filter example, and the MCS6502.

Blackburn88

Robert L. Blackburn, Donald E. Thomas, and Patti M. Koenig, "CORAL II: Linking Behavior and Structure in an IC Design System", *Proc. of the 25th DAC*, pages 529–535, June 1988.

Linking values and operators in the GDB (ISPS parse tree), the Value Trace, the allocated hardware, and the control step schedule.

Walker88

Robert Allen Walker, *Design Representation and Behavioral Transformation for Algorithmic Level Integrated Circuit Design*, PhD Thesis, Dept. of Electrical and Computer Engineering, Carnegie Mellon University, April 1988.

A model of design representation, algorithmic level behavioral and structural transformations, adding concurrent processes to a fifth-order digital elliptic wave filter example, pipelining the MCS6502 and the RISC-1.

Koenig87

Patti M. Koenig, *SEE-SAW: A Graphical Interface for System Level Design*, Master's Thesis, Dept. of Electrical and Computer Engineering, Carnegie Mellon University, December 1987.

A graphical interface for viewing links between the ISPS behavioral description, the synthesized structure, and the control step schedule.

Walker87

Robert A. Walker and Donald E. Thomas, "Design Representation and Transformation in the System Architect's Workbench", *Proc. of ICCAD'87*, pages 166–169, November 1987.

Algorithmic level behavioral and structural transformations, and pipelining the MCS6502.

Nestor87

John Anthony Nestor, *Specification and Synthesis of Digital Systems with Interfaces*, PhD Thesis, Dept. of Electrical and Computer Engineering, Carnegie Mellon University, April 19887.

Behavioral specification with interfaces, synthesis with interfaces, control step scheduling, the Intel 8251, a Multibus example, a protocol adapter example, and a block transfer example.

Nestor86

J.A. Nestor and D.E. Thomas, "Behavioral Synthesis with Interfaces", *Proc. of ICCAD'86*, pages 112–115, November 1986.

Behavioral specification with interfaces, synthesis with interfaces, control step scheduling, and a block transfer example.

Walker85

Robert A. Walker and Donald E. Thomas, "A Model of Design Representation and Synthesis", *Proc. of the 22nd DAC*, pages 453–459, June 1985.

A model of design representation with three domains of description (Behavioral, Structural, and Physical) and multiple levels of abstraction.

Rajan85

Jayanth V. Rajan and Donald E. Thomas, "Synthesis by Delayed Binding of Decisions", *Proc. of the 22nd DAC*, pages 367–373, June 1985.

An older version of SUGAR.

Carnegie Mellon's Facet / Emerald System

The Facet design methodology, and its implementation as the Emerald system (actually Emerald I and Emerald II), was Tseng's thesis work at CMU. Facet/Emerald includes scheduling and data path synthesis. See also *AT&T's Bridge System*, Tseng's continuation of this work at AT&T Bell Labs.

Input

ISPS.

Internal Behavioral Representation

Carnegie Mellon's Value Trace (VT), a dataflow / controlflow graph.

Scheduling

Uses ASAP scheduling, calling it AEAP (As Early As Possible), and applying it to one basic block at a time.

Data Path Synthesis

Register Binding

First, builds a compatibility graph, with one node per variable, and an edge connecting variables that can be assigned to the same register (variables that are live in different control steps). Then finds the minimum number of cliques, thus minimizing the number of registers.

Uses as many ALUs as necessary.

Functional Unit Binding

First builds a compatibility graph, with one node per data operator, and an edge connecting operators that can be assigned to the same ALU (operators not used in the same control step). Then finds the minimum number of cliques, thus minimizing the number of ALUs. As the cliques are found, the algorithm gives priority to grouping together operators with (1) the same inputs and outputs, and (2) performing the same operation.

Interconnection

First builds a compatibility graph, with one node per interconnection variable, and an edge connecting variables that can never be used simultaneously. Then finds the minimum number of cliques, thus minimizing the the number of interconnection units.

Examples

The AM2910, AM2901, MCS6502, and IBM System/370.

References

Tseng86b

Chia-Jeng Tseng and Daniel P. Siewiorek, "Automated Synthesis of Data Paths in Digital Systems", *IEEE Trans. on CAD*, pages 379–395, July 1986.

Scheduling, data path synthesis, design space exploration, the AM2910, the AM2901, and the IBM System/370. Essentially Tseng83 + Tseng84a.

Tseng84a

Chia-Jeng Tseng and Daniel P. Siewiorek, "Emerald: A Bus Style Designer", *Proc. of the 21st DAC*, pages 315–321, June 1984.

Design space exploration and results for the AM2910 and the AM2901.

Tseng84b

Chia-Jeng Tseng, *Automated Synthesis of Data Paths in Digital Systems*, PhD Thesis, ECE Dept., CMU, April 1984.

Scheduling, data path synthesis, design space exploration, the AM2910, the AM2901, the MCS6502, and the IBM System/370.

Tseng83

Chia-Jeng Tseng and Daniel P. Siewiorek, "Facet: A Procedure for the Automated Synthesis of Digital Systems", *Proc. of the 20th DAC*, pages 490–496, June 1983.

Scheduling and data path synthesis.

Tseng81

Chia-Jeng Tseng and Daniel P. Siewiorek, "The Modeling and Synthesis of Bus Systems", *Proc. of the 18th DAC*, pages 471–478, June 1981.

Data path synthesis.

Case Western's ADPS System

Case Western Reserve University's ADPS (A Data Path Synthesizer) system includes scheduling and data path synthesis.

Input

Not mentioned in the literature.

Internal Behavioral Representation

A data flow graph.

Scheduling / Functional Unit Allocation

A linear programming algorithm is used to schedule the operations into control steps and allocate the number of functional units of each type (including single-function and multiple-function units), while minimizing the total ALU cost.

Data Path Synthesis

Heuristics are used to bind the operations to functional unit instances, and a clique partitioning algorithm is used to allocate and bind registers. Finally, multiplexors and interconnections are added in a straight-forward manner, and optimized using a set of interconnect transformations.

Examples

A sixth-order elliptic band-pass filter example, a complex biquad recursive digital filter example, a fifth-order digital elliptic wave filter example, a second-order differential equation example, and Tseng's running example.

References

Papachristou90

C.A. Papachristou and H. Konuk, "A Linear Program Drive Scheduling and Allocation Method Followed by an Interconnect

Optimization Algorithm", *Proc. of the 27th DAC*, pages 77–83, June 1990.

Scheduling and data path synthesis, a sixth-order elliptic band-pass filter example, a complex biquad recursive digital filter example, a fifth-order digital elliptic wave filter example, a second-order differential equation example, and Tseng's running example.

Eindhoven Univ. of Technology's <Esc> System

Eindhoven University of Technology's <Esc> (Eindhoven Silicon Compiler) system performs scheduling and data path synthesis (EASY —Eindhoven Architecture Synthesis), logic synthesis, and layout synthesis. Several synthesis techniques have been explored.

Input

A Pascal- or C-like language.

Internal Behavioral Representation

A demand graph, a form of dataflow / controlflow graph.

Transformations

Code motion, variable folding, dead code elimination, etc.

Scheduling

Stok88

Schedules one basic block at a time, using a variation of force-directed scheduling. First, the possible control steps are determined, using ASAP and ALAP scheduling. The algorithm then distributes the operations over the control steps so as to balance the use of each functional unit, taking into account the probability that the operation will be placed into each control step, as well as favoring implementation by cheaper modules. Multi-cycle operations and chaining are supported.

Jess88

Initially, all operations are assigned to a single control step. This step is then split due to cycles, branches, procedure calls, and when the chaining exceeds the length of the step or the area exceeds the area constraint.

Data Path Synthesis

Stok90

Variables are merged into register files using an edge coloring algorithm. If a two-phase clocking scheme is used, the states can be split into a write (from the register file) part and a read (into the register file) part, the nodes in the state graph can be duplicated, and a bipartite edge coloring algorithm can be used to merge the variables. The number of registers in the file can then be determined by lifetime analysis, using a clique partitioning algorithm or the left edge algorithm to find the minimum number of registers and their allocation.

Operations are assigned to functional units as in Stok88. As this is done, the local connections from the functional unit to the buses are optimized using a simulated annealing algorithm.

Stok88

First, operations are assigned to functional units, using a weighted clique partitioning algorithm that takes into account the advantage (or disadvantage) of combining two operations into a single functional unit, as well as the multiplexor cost.

Registers are then allocated by performing lifetime analysis, building a compatibility graph to represent values that are not alive simultaneously, and using the left edge algorithm to find the maximum clique cover, and thus the minimum number of registers and their allocation.

Finally, registers are merged into register files, which reduces the number of buses. Merging is allowed only if the registers are not written or read in the same cycle. Again, a weighted clique partitioning algorithms used.

Jess88

Registers are assigned using the left edge algorithm, as above.

Operations are assigned using a weighted compatibility graph, with the nodes (operations) connected by the lowest-cost edge merged, and this process repeated until no more edges are present.

Registers are combined into register files using an interval graph, grouping the registers with the highest cost first.

These three tasks can also be performed simultaneously, using a dynamic programming method.

Examples

A fifth-order digital elliptic wave filter example and Tseng's running example.

References

Stok90

L. Stok, "Interconnect Optimisation During Data Path Allocation", *Proc. of EDAC'90*, pages 141–145, March 1990.

Data path synthesis, and a fifth-order digital elliptic wave filter example.

Stok88

L. Stok and R. van den Born, "EASY: Multiprocessor Architecture Optimisation", in *Logic and Architecture Synthesis for Silicon Compilers (Proc. of the Int. Workshop on Logic and Arch. Synth. for Silicon Compilers)*, G. Saucier and P.M. McLellan (Editors), pages 313–328, Elsevier Science Publishers, May 1988.

<Esc> and EASY overview, scheduling, data path synthesis, and a fifth-order digital elliptic wave filter example.

Jess88

J.A.G. Jess, R. v.d. Born, and L. Stok, "Synthesis of Concurrent Hardware Structures", *Proc. of ISCAS'88*, pages 2757–2760, June 1988.

The demand graph, transformations, scheduling, data path synthesis, and Tseng's running example.

GE's FACE / PISYN System

GE's FACE (Flexible Architecture Compilation Environment) Core Environment provides a common model, set of graph algorithms, set of development tools, and framework for synthesis tools. The PISYN toolset, for synthesizing pipelined systems, includes data path synthesis, scheduling, module binding and analysis. PISYN is targeted for DSP systems (pipelined architecture), although some of the algorithms are generic and useful for many architectures.

Input

FACE includes a parser for SMALL, a GE-designed and implemented hardware design language, and includes an interactive graph editor for graphical entry.

Internal Behavioral Representation

Uses a dataflow hypergraph (a hypergraph consists of a set of vertices, edges, and relations between vertices and edges), with operators mapped onto vertices, values mapped onto edges, and relations joining vertices and edges. The same representation is used for structural information.

Data Path Synthesis

Maps operations onto functional units, and minimizes the total execution time of expressions using associativity and commutativity. Shares functional units among mutually exclusive conditional branches, inserting multiplexors as necessary.

First Scheduling Option

This option is for a pipeline where data enters the pipeline every clock cycle, and operators have been previously configured (i.e., time vs. area implementations for the operators have been selected). The designer specifies the system clock time.

Stage Scheduling

Schedules the functional units into stages and inserts stage registers, minimizing the number of pipeline stages in the design.

Second Scheduling Option

This option is for a pipeline where data enters the pipeline less often than every clock cycle (i.e., when the system clock is faster than the data rate). Hardware is shared within the pipeline since hardware can be time multiplexed using the faster clock. This algorithm is applied to a design that has been configured earlier. The designer specifies the system clock time and the desired number of stages in the pipeline.

Stage Scheduling

First, ASAP and ALAP scheduling is used to determine the freedom (similar to mobility) of each operation. Then the design is scheduled to maximize the potential hardware sharing: for each type of operation, from the most to the least expensive, the algorithm iteratively schedules the least free operation in the least dense partition (for a data introduction interval of n, there are n partitions). Then operations in different partitions are bound to the same functional unit to achieve hardware sharing and to simultaneously attempt to minimize the multiplexing cost. This algorithm usually has the effect of inserting gaps into the pipeline schedule (NO-OP stages) to improve hardware sharing between stages.

Third Scheduling Option

This option is used when operator configuration is desired in addition to minimum latency scheduling. Data is assumed to arrive at the system clock rate. The user specifies the system clock time and the desired number of stages. For designs where operator configurations are known or specified directly by the user, the first scheduling option is much more efficient.

Stage Scheduling

In the most current version, an upper bound heuristic is used to bound the possible area and time solutions at each stage (corresponding to an operator in the topologically sorted data flow graph). A dynamic programming algorithm simultaneously selects the optimal operator configurations and stages for the design in one step, inserting necessary stage latches into one of the optimal designs.

Fourth Scheduling Option

This option is a combination of the second and third scheduling options. The second scheduling option is run using minimum time operators to allow maximum freedom of scheduling and sharing of hardware. Then the third scheduling option is used with constraints of jammed pipeline stages and shared hardware calculated using the second scheduling option.

Examples

A large DSP example from a computer image generation application, a Chebychev preprocessor, a pipelined 16-point FIR filter example, a pipelined fifth-order digital elliptic filter example, and various smaller examples.

References

McNall90

Kristen N. McNall and Albert E. Casavant, "Automatic Operator Configuration in the Synthesis of Pipelined Architectures", *Proc. of the 27th DAC*, pages 174–179, June 1990.

Control step scheduling, data path synthesis, a small example, and a Chebyshev approximator example.

Hwang89

Ki Soo Hwang, Albert E. Casavant, Ching-Tang Chang, and Manuel A. d'Abreu, "Scheduling and Hardware Sharing in Pipelined Data Paths", *Proc. of ICCAD'89*, pages 24–27, November 1989.

Pipeline scheduling and hardware sharing, a pipelined 16-point FIR filter example, and a pipelined fifth-order digital elliptic filter example.

Hartley89

Richard Hartley and Albert E. Casavant, "Tree-Height Minimization in Pipelined Architectures", *Proc. of ICCAD'89*, pages 112–115, November 1989.

Minimizing the execution time of expressions by reducing the height of trees of commutative and associative operators.

Smith89

W.D. Smith, D. Duff, M. Dragomirecky, J. Caldwell, M. Hartman, J. Jasica, and M.A. d'Abreu, "FACE Core Environment: The Model and its Application in CAE/CAD Tool Development", *Proc. of the 26th DAC*, pages 466–471, June 1989.

The FACE Core Environment, including the behavioral and structural representation.

Dragomirecky89

Martin Dragomirecky, Ephraim P. Glinert, Jeffrey R. Jasica, David A. Duff, William D. Smith, and Manuel A. d'Abreu, "High-Level Graphical User Interface Management in the FACE Synthesis Environment", *Proc. of the 26th DAC*, pages 549–554, June 1989.

VISAGE, FACE's object-oriented user interface framework.

Casavant89

A.E. Casavant, M.A. d'Abreu, M. Dragomirecky, D.A. Duff, J.R. Jasica, M.J. Hartman, K.S. Hwang, and W.D. Smith, "A Synthesis Environment for Designing DSP Systems", *IEEE Design and Test*, pages 35–44, April 1989.

DSP synthesis, including transformations, data path synthesis, and a pipelined DSP example.

Hwang88

Ki Soo Hwang, Albert E. Casavant, Martin Dragomirecky, and Manuel A. d'Abreu, "Constrained Conditional Resource Sharing in Pipeline Synthesis", *Proc. of ICCAD'88*, pages 52–55, November 1988.

Sharing functional units between mutually exclusive conditional branches.

Smith88

William D. Smith, Jeffrey R. Jasica, Michael J. Hartman, and Manuel A. d'Abreu, "Flexible Module Generation in the FACE Design Environment", *Proc. of ICCAD'88*, pages 396–399, November 1988.

Module generation using parameterized, procedural module descriptions.

Honeywell's V-Synth System

Honeywell's V-Synth (VHDL-Synthesis) System decomposes VHDL descriptions into simple statements, optimizes those statements with compiler-like transformations, schedules the operations into states, and then passes the results to the MIMOLA Synthesis System to complete the design process. See also *Univ. of Kiel's MIMOLA System*.

Input

VHDL behavioral descriptions.

Internal Behavioral Representation

Uses a process graph, which represents the behavior as a set of nodes, or blocks, connected by directed arcs. Each block contains a set of statements, with each statement either a simple assignment statement (one operation, and at most two operators), or a simple (IF...THEN...ELSE) conditional statement. If a conditional statement is included, it must be the last statement in the block.

Transformations

Compiler-like transformations, including constant folding, code motion, global common-subexpression elimination, expression factoring, and redundant code elimination.

Scheduling

Uses back-substitution to rewrite the statements in each block into a form where each assignment statement has an output variable on the left-hand side, and variables other than output variables on the right-hand side. The resulting statements can then be executed in any sequence, or in parallel. Once this potential parallelism is identified, the designer controls whether or not the parallelism is exploited.

Data Path Synthesis

Allocates registers and memories to variables, and uses a clique partitioning algorithm to allow sharing. This clique partitioning algorithm gives priority to the least-connected vertices.

Uses the MIMOLA Synthesis System to complete the design, assigning functional units and interconnection.

Examples

A bubble sort example, and six code fragment examples.

References

Bhasker90

J. Bhasker and Huan-Chih Lee, "An Optimizer for Hardware Synthesis", *IEEE Design and Test*, pages 20–36, October 1990.

V-Synth overview, generating the process graph, transformations, scheduling, and six code fragment examples.

Harper89

Paul Harper, Stanley Krolikoski, and Oz Levia, "Using VHDL as a Synthesis Language in the Honeywell Vsynth System", in *Computer Hardware Description Languages and their Applications (Proc. of the 9th Int. Conf. on CHDLs)*, John A. Darringer and Franz J. Rammig (Editors), pages 315–330, North Holland, June 1989.

V-Synth overview, the process graph model, and the behavioral/ structural model used as an input to MIMOLA.

Bhasker88a

J. Bhasker and Tariq Samad, "A Better Clique-Partitioning Algorithm", *Proc. of the Allerton Conf. on Communication, Control, and Computing*, September 1988.

Clique partitioning.

Bhasker88b

J. Bhasker, "Process-Graph Analyser: A Front-End Tool for VHDL Behavioural Synthesis", *Software — Practice and Experience*, pages 469–483, May 1988.

V-Synth overview, generating the process graph, transformations, and scheduling.

Bhasker88c

J. Bhasker, "Implementation of an Optimizing Compiler for VHDL", *SIGPLAN Notices*, January 1988.

Parsing a VHDL description to produce the process graph, and transformations.

Bhasker87a

J. Bhasker, "An Algorithm for Microcode Compaction of VHDL Behavioral Descriptions", *Proc. of Micro-20*, December 1987.

The process graph model, scheduling, and a bubble sort example.

Bhasker87b

J. Bhasker and Tariq Samad, "Compacting MIMOLA Microcode", *Proc. of Micro-20*, December 1987.

Using a clique-partitioning algorithm to recode MIMOLA's horizontal microinstructions as vertical microinstructions.

Krolikoski86

Stanley J. Krolikoski, Jayaram Bhasker, and Subra Natarajan, "The V-Synth System", *Proc. of the Honeywell CAD/CAM Conf.*, 1986.

V-Synth overview.

IBM's ALERT System

IBM's ALERT system was one of the first high-level synthesis systems, and included scheduling, data path synthesis, and structural transformations. The output was a Register-Transfer level design, specified in terms of Boolean equations.

Input

A language based on Iverson notation (used in APL). The language supports arithmetic and logical operations, assignment, branching, arrays, macros, etc. The behavioral description describes storage, registers, and any predefined pieces of logic, such as adders or decoders.

Internal Behavioral / Structural Representation

A dataflow / controlflow design file, representing the initial structure of the design.

Variable subscripting (array references) is replaced with logic to select the appropriate element, and macros are expanded.

Scheduling

The user is allowed to prespecify all or part of the schedule.

Initially, all operations are assigned to separate a single control step. This step is then split at statements that are destinations of GOTOs, after conditional branch statements, and whenever a variable receives a second value in a given control step.

Data Path Synthesis

Dataflow analysis is used to assign values to flip-flops (or latches), storing those values that are produced in one control step and used in a later control step.

Transformations

Performs common subexpression elimination, and substitutes logically equivalent structures when doing so will reduce the number of components.

Examples

A code fragment for loading an index register, and an IBM 1800 computer.

References

Friedman70

Theodore D. Friedman and Sih-Chin Yang, "Quality of Designs from an Automatic Logic Generator (ALERT)", *Proc. of the 7th Design Automation Workshop*, pages 71–89, June 1970.

ALERT overview, design and analysis of an IBM 1800 computer.

Friedman69

Theodore D. Friedman and Sih-Chin Yang, "Methods Used in an Automatic Logic Design Generator (ALERT)", *IEEE Trans. on Computers*, pages 593–614, July 1969.

ALERT overview, input language based on Iverson notation, scheduling, data path synthesis, transformations, and a code fragment for loading an index register.

IBM's HIS System

IBM's HIS (High-Level IBM Synthesis) system includes scheduling and data path synthesis. See also IBM's *Yorktown Silicon Compiler* and *Univ. of Karlsruhe's DSL Synthesis System / CADDY System* — Camposano was previously involved with those systems.

Input

VHDL behavioral descriptions (procedures).

Also reads a text form of YIF (Yorktown Internal Format). See *IBM's Yorktown Silicon Compiler.*

Internal Behavioral Representation

Uses a representation called the SSIM (Sequential Synthesis Incore Model). The SSIM contains linked dataflow and controlflow graphs, and is hierarchical, in that it can represent several modules at once, and each can be synthesized separately.

Dataflow Analysis / Scheduling

Camposano90a, Camposano90b

As Fast as Possible (AFAP) path-based scheduling: First, all possible execution paths through the controlflow graph are found, and timing and resource constraints are expressed as intervals where the paths must be cut to separate the operations into states. Clique partitioning is then used on each path to find the minimum number of cuts which meet the constraints, resulting in the minimum number of control steps for each path. Finally, the schedules for all paths are merged to form the final control step schedule, using clique partitioning to find the minimum number of states for all paths.

Camposano90a

Heuristic path-based scheduling: First, all possible execution paths through the controlflow graph are found, and timing and resource constraints are expressed as intervals where the paths must be cut to separate the operations into states. Then heuristics are used to compute a cutting point, and cut all paths containing that point. The heuristics perform recursive backtracking, evaluating all cuts in a path and choosing the one with the minimum cost. Cut-early, cut-late, and look-ahead versions of these heuristics were implemented.

Bergamaschi91b

Since each variable can hold different values at different times, dataflow analysis is used to determine the lifetime of each value. This dataflow analysis is global, not restricted to individual basic blocks.

Uses AFAP path-based scheduling, as described above. After the paths are merged, where there is more than one possible cut, the dataflow analysis is used to choose the cut with the smallest number of live values, thus minimizing the number of registers.

During scheduling, if an operation can be implemented by more than one type of functional unit, it is assigned to a functional unit using a constraint-interval-based algorithm, giving priority to operations that can be assigned to fewer functional units, and attempting to balance functional unit usage.

Data Path Synthesis

An initial data path is generated, using one single-function functional unit for each operation in each path segment, using registers to store values generated in one control step and used in another (storing all values assigned to the same variable in the same register), and interconnecting those components with wires and multiplexors. Control information is also generated during this phase.

The initial data path is then optimized, coloring conflict graphs to minimize the number of functional units, registers, and multiplexors.

Examples

The Intel 8251, the MCS6502, the FRISC microprocessor, a Kalman filter example, a fifth-order digital elliptic filter example, a parity generator example, a greatest common divisor algorithm example, a counter example, MAHA's example, a prefetch buffer example, a second-order differential equation example, a differential equation solver, an arithmetic circuit example, and a traffic light controller example.

References

Bergamaschi91a

Reinaldo A. Bergamaschi, Raul Camposano and Michael Payer, "Data-Path Synthesis using Path Analysis", *Proc. of the 28th DAC,* June 1991.

Data path synthesis and optimization.

Camposano91a

R. Camposano, R.A. Bergamaschi, C.E. Haynes, M. Payer, and S.M. Wu, "The IBM High-Level Synthesis System", in *Trends in High-Level Synthesis*, R. Camposano and Wayne Wolf (Editors), Kluwer, 1991.

HIS overview, SSIM, dataflow analysis, AFAP scheduling, module assignment, path-based allocation, the Intel 8251, the MCS6502, the FRISC microprocessor, a prefetch buffer example, a Kalman filter example, a fifth-order digital elliptic filter example, a differential equation solver, an arithmetic circuit example, a traffic light controller example, a greatest common divisor algorithm example, and a counter example.

Bergamaschi91b

Reinaldo A. Bergamaschi, Raul Camposano and Michael Payer, "Area and Performance Optimizations in Path-Based Scheduling", *Proc. of EDAC'91*, February 1991.

Functional unit assignment, register minimization, MAHA's example, a second-order differential equation example, and the FRISC microprocessor.

Camposano91b

Raul Camposano, "Path-Based Scheduling for Synthesis", *IEEE Trans. on CAD*, pages 85–93, January 1991.

AFAP path-based scheduling, the Intel 8251, a Kalman filter example, a greatest common divisor algorithm example, a counter example, and others. Essentially an updated version of Camposano90a.

Camposano90a

Raul Camposano and Reinaldo A. Bergamaschi, "Synthesis Using Path-Based Scheduling: Algorithms and Exercises", *Proc. of the 27th DAC*, pages 450–455, June 1990.

AFAP path-based scheduling, heuristic path-based scheduling, the Intel 8251, a parity generator example, the MCS6502, a Kalman filter example, a greatest common divisor algorithm example, a counter example, and others.

Camposano90b

R. Camposano, "Path-Based Scheduling for Synthesis", *Proc. of the Hawaiian Int. Conf. on System Sciences*, pages 348–355, January 1990.

AFAP path-based scheduling, the Intel 8251, a Kalman filter example, a greatest common divisor algorithm example, a counter example, and others.

Camposano89

R. Camposano and R.M. Tabet, "Design Representation for the Synthesis of Behavioral VHDL Models", in *Computer Hardware Description Languages and their Applications (Proc. of the 9th Int. Conf. on CHDLs)*, John A. Darringer and Franz J. Rammig (Editors), pages 4958–205, North Holland, June 1989.

Representing a design as linked dataflow, controlflow, data path, and control automation graphs; and the scheduling and allocation process.

IBM's V Compiler

IBM's V compiler produces Register- Transfer level designs, which can then passed on to the LSS system for logic synthesis. The V compiler includes scheduling and data path synthesis, and can produce high level software simulators for the designs. A subset of the language is used to produce YIF format output for IBM's Yorktown Silicon Compiler and IBM's HIS System.

Input

The V language — a procedural, C-like language. V includes tasks (similar to processes), asynchronous calls, and queues.

V also includes a cycle-block construct, which can specify how certain references to communication ports should be scheduled relative to each other. Other operations are free to be scheduled before, during, or after the block, as appropriate.

Internal Behavioral Representation

Each task of the parse tree, with the leaves representing source tokens and vertices representing semantic units, is mapped into a graph incorporating both control flow and data flow, with vertices representing primitive operations. Lifetime analysis is used to represent the data flow in single-assignment form.

Scheduling

The controlflow/dataflow graphs are rearranged and optimized. Using a percolation algorithm, the vertices are scheduled into specific states, subject to data dependencies, resource limits, and other constraints.

Data Path Synthesis

The data flow graph is inspected, and flows which cross cycle boundaries are assigned to registers using a graph coloring algorithm. Operations are also assigned to functional units during this phase. Multiplexors are assigned as required for the registers and functional units.

Examples

A greatest common divisor example, and the FRISC microprocessor.

References

Berstis89a

Viktors Berstis, "Cycle-Level Timing Constraints in Behavioral Hardware Descriptions", in *Computer Hardware Description Languages and their Applications (Proc. of the 9th Int. Conf. on CHDLs)*, John A. Darringer and Franz J. Rammig (Editors), pages 197–205, North Holland, June 1989.

Cycle-blocks, scheduling, and a simple microprocessor example.

Berstis89b

Viktors Berstis, "The V Compiler: Automating Hardware Design", *IEEE Design and Test*, pages 8–17, April 1989.

The V language, and IBM's V Compiler.

Berstis85

Viktors Berstis, Daniel Brand, and Ravi Nair, "An Experiment in Silicon Compilation", *Proc. of ISCAS'85*, pages 655–658, June 1985.

The V language, logic synthesis, and the FRISC microprocessor.

IBM's Yorktown Silicon Compiler

IBM's Yorktown Silicon Compiler includes partitioning, scheduling, data path synthesis, design iteration, controller design, logic synthesis, and timing optimization. See also *Univ. of Karlsruhe's DSL Synthesis System / CADDY System* — Camposano was previously involved with that system, and *IBM's HIS System* — Camposano is now involved with that system.

Input

The V language, a procedural, Pascal-like language. V includes tasks (similar to processes), asynchronous calls, and queues.

Also reads a text form of YIF (Yorktown Internal Format).

Internal Behavioral Representation

Uses a dataflow graph and a controlflow graph, described in the Yorktown Internal Format (YIF). In the dataflow graph, the nodes represent operations and variables, and the arcs represent data dependencies. In the controlflow graph, the nodes represent operations, and the arcs represent predecessor / successor relationships.

Scheduling / Register Allocation

Camposano90

Heuristic path-based scheduling: First, all possible execution paths through the controlflow graph are found, and timing and resource constraints are expressed as intervals where the paths must be cut to separate the operations into states. Then heuristics are used to compute a cutting point, and cut all paths containing that point.

Camposano89, Camposano88, Brayton88, Camposano87d, Camposano87e, Brayton86

Starts with a single control state, then splits that state as follows. Additional states are added to permit looping and procedure calls (called "module calls"). States are also added due to timing constraints and data flow restrictions. When a value written in one state is read in another state, a register is added to hold the value.

States can be split further or folded to meet constraints.

Design Iteration

First, a control step schedule with the minimum number of control steps is generated, using one of the schedulers above. Then the execution paths in each control step are traversed, and the area required for the data operations is estimated. Finally, the path / control step with the largest area is selected for splitting, and is split into two control steps so that the areas of the two subpaths is approximately equal. By continuing this process, an area-time tradeoff curve can be generated.

Data Path Synthesis

Each operation is initially implemented by a separate (combinational) functional unit, and each variable by a separate register, producing blocks of combinational logic (one block per procedure) and interconnection. For each block of combinational logic, a clique partitioning algorithm is then used to fold together functional units performing the same operation in different time steps or branch alternatives, and to fold together registers as guided by lifetime analysis.

Controller Design

Control signals are identified, and combinational logic is synthesized to produce those signals.

Partitioning

Partitioning a design into smaller blocks results in smaller, more manageable blocks, and in a faster logic synthesis phase. Both manual and automatic partitioning are supported.

Since the communication between procedures maps directly into port-oriented communication between blocks, a design can be manually partitioned using procedures.

A design can also be partitioned into smaller blocks using a clustering technique. Two different clustering algorithms are given in Camposano87a. The first considers only the connections between the clusters and their area (tending toward clusters of equal size with a small number of interconnections), while the second considers the similarity of the operations as well.

Logic Synthesis / Timing Optimization

The Yorktown Logic Editor (YLE) is used to synthesize a network of combinational blocks. If this network is too large to be handled, it must be partitioned as described earlier. Note that a network can contain both data and control logic.

For each block of combinational logic, a logic specification is generated and minimized (for area) to produce a multi-level logic implementation. Timing optimization then finds the critical path and reduces delay using transistor resizing, logic resynthesis, etc.

Examples

A 4-stage IBM 801, a greatest common divisor algorithm example, a traffic light controller, and an IBM System/370 EDIT instruction example.

References

Camposano90

Raul Camposano and Reinaldo A. Bergamaschi, "Redesign Using State Splitting", *Proc. of EDAC'90*, pages 157–161, March 1990.

Heuristic AFAP scheduling, design iteration, the Intel 8251, a Kalman filter example, the FRISC, and a greatest common divisor algorithm example.

Camposano89

R. Camposano, "Behavior-Preserving Transformations for High-Level Synthesis", *Proc. Workshop on Hardware Specification, Verification, and Synthesis: Mathematical Aspects*, Springer-Verlag, 1989.

Correctness proofs for scheduling and data path synthesis transformations, and an IBM System/370 EDIT instruction example.

Camposano88

Raul Camposano, "Design Process Model in the Yorktown Silicon Compiler", *Proc. of the 25th DAC*, pages 489–494, June 1988.

The YIF internal representation, scheduling, data path synthesis, logic synthesis and timing optimization, and a 4-stage IBM 801.

Brayton88

R.K. Brayton, R. Camposano, G. De Micheli, R.H.J.M. Otten, and J. van Eijndhoven, "The Yorktown Silicon Compiler System", in *Silicon Compilation*, Daniel D. Gajski (Editor), pages 204–310, Addison-Wesley, 1988.

Same as Brayton86.

Camposano87a

R. Camposano and R.K. Brayton, "Partitioning Before Logic Synthesis", *Proc. of ICCAD'87*, pages 324–326, November 1987.

Two partitioning algorithms and several unknown examples.

Camposano87b

R. Camposano and J.T.J. van Eijndhoven, "Combined Synthesis of Control Logic and Data Path", *Proc. of ICCAD'87*, pages 327–329, November 1987.

Controller design and logic synthesis and several unknown examples.

Camposano87c

R. Camposano and J.T.J. van Eijndhoven, "Partitioning a Design in Structural Synthesis", *Proc. of ICCD'87*, pages 564–566, October 1987.

Manual and automatic partitioning and the IBM 801 (called a 32-bit microprocessor).

DeMicheli87

Giovanni De Micheli, "Performance-Oriented Synthesis of Large-Scale Domino CMOS Circuits", *IEEE Trans. on CAD*, pages 751–765, September 1987.

Timing performance optimization, example using a 32-bit microprocessor (a 4-stage IBM 801).

Camposano87d

Raul Camposano, "Structural Synthesis in the Yorktown Silicon Compiler", *Proc. of VLSI'87*, pages 61–72, August 1987.

Control step scheduling, data path synthesis, controller design, a greatest common divisor algorithm example, and a traffic light controller.

Camposano87e

R. Camposano, *A Method for Structural Synthesis*, IBM Research Report RC 13081, August 1987.

Design representation, control step scheduling, data path synthesis, and an IBM System/370 EDIT instruction example. Early version of Camposano89.

Brayton86

R.K. Brayton, R. Camposano, G. De Micheli, R.H.J.M. Otten, and J. van Eijndhoven, *The Yorktown Silicon Compiler System*, IBM Research Report RC 12500, February 1986.

History, design representation, control step scheduling, data path synthesis, controller generation, logic synthesis, layout design, timing optimization, and the IBM 801. Very detailed.

Brayton85a

R.K. Brayton, C.L. Chen, G. De Micheli, and R.H.J.M. Otten, "A Microprocessor Design Using the Yorktown Silicon Compiler", *Proc. of ICCD'85*, pages 225–231, October 1985.

A 4-stage IBM 801.

Brayton85b

R.K. Brayton, N.L. Brenner, C.L. Chen, G. DeMicheli, C.T. McMullen, and R.H.J.M. Otten, "The Yorktown Silicon Compiler", *Proc. of ISCAS'85*, pages 391–394, June 1985.

Logic synthesis and layout design.

IIT Delhi's Synthesis System

IIT Delhi's synthesis system includes design iteration, scheduling and data path synthesis. See also *Univ. of Kiel's MIMOLA System* — Marwedel is also involved with that system.

Input

A subset of the MIMOLA language, a Pascal-like language that includes recursive procedure calls and multi-dimensional arrays.

Internal Behavioral Representation

A dataflow graph.

Design Iteration

The system chooses a set of operators to evaluate, based on data dependencies and functional unit availability. These operators are scheduled and bound, and another set of operators is chosen for the next iteration. Note that this also allows a partially-specified design to be completed by the synthesis system.

Scheduling

List scheduling, with three scheduling modes: forward scheduling (using ASAP), backward scheduling (using ALAP), and double-headed scheduling (alternating between ASAP and ALAP). For forward scheduling, the set of data-ready operations are scheduled into the current control step, with operations on the longest path scheduled first.

Data Path Synthesis

Operations are bound to functional units, and values are bound to registers, using a zero-one integer programming method, and attempting to minimize the amount of interconnection. This method can also perform tradeoffs between different-speed functional units that perform the same operation.

Registers are combined into multiport memories, using a zero-one linear programming method, which considers the access requirements of the registers, as well as their interconnection to operators.

Examples

A differential equation example, a digital filter example, and a fifth-order digital elliptic filter example.

References

Balakrishnan89

M. Balakrishnan and P. Marwedel, "Integrated Scheduling and Binding: A Synthesis Approach for Design Space Exploration", *Proc. of the 26th DAC*, pages 68–74, June 1989.

Iteration, scheduling, data path synthesis, a differential equation example, a digital filter example, and a fifth-order digital elliptic filter example.

Balakrishnan88

M. Balakrishnan, Arun K. Majumdar, Dilip K. Banerji, James G. Linders, and Jayanti C. Majithia, "Allocation of Multiport Memories in Data Path Synthesis", *IEEE Trans. on CAD*, pages 536–540, April 1988.

Grouping registers into multiport memories.

IMEC's CATHEDRAL-II System

IMEC's CATHEDRAL-II system is a design-style-specific system for the synthesis of digital signal processing (DSP) chips. CATHEDRAL-II includes transformations, data path synthesis, scheduling, and controller design. See also *IMEC's CATHEDRAL-2nd System*, *IMEC's CATHEDRAL-III System*, and *Philip's PIRAMID System*, which uses the CATHEDRAL-II system as its synthesis engine.

Input

SILAGE, a Pascal-like language designed for describing DSP algorithms. SILAGE supports time domain operations, such as interpolation, and allows the designer to pass structural "hints" to the synthesis tools.

Internal Behavioral Representation

SILAGE primitives.

Transformations

Performs constant folding and common subexpression elimination.

Data Path Allocation — Jack-the-Mapper

A CATHEDRAL-II-generated data path is built from a set of six execution units (EXUs), as well as a set of memories, I/O units, and controller modules. The EXUs include an ALU / shift unit, an address computation unit, a parallel multiplier / accumulator, a parallel / serial divider, a comparator, and a normalizer-scaler, and are composed of adders, shifters, etc.

First, the designer specifies the number of EXUs of each type. Then, during the data path allocation phase, a partially-rule-based, partially-algorithmic translator assigns each operation to a particular type of EXU, assigns variables to register files, and generates a dedicated bus for each variable to transfer it between the necessary EXUs and register files. However, each operation is not assigned to a specific EXU during this phase, and each variable is not assigned to a specific location in the register file.

Scheduling / EXU Assignment — Atomics

Performs loop folding, overlapping the execution of successive loop iterations.

List scheduling, with priority to operations on the longer critical paths.

A vertex coloring technique is used to assign operations to specific EXUs.

Data Path Assignment

Values are bound to specific registers of the register file based on lifetime analysis. For a specific register, values are packed into it throughout its lifetime in a greedy fashion. If values remain, this process is repeated with a new register, continuing until all values have been assigned.

A bus-merging algorithm is used to minimize the number of buses. Beginning with i = 1, the buses are ordered by usage, the i most-used buses are selected, and the other buses are merged in if possible. If not, i is incremented and the ordering, selecting, and merging process continues.

Examples

A 256-point DFT-algorithm example, and a fifth-order filter for a PCM telephony application.

References

Goossens90

Gert Goossens, Jan Rabaey, Joos Vandewalle, and Hugo De Man, "An Efficient Microcode Compiler for Application Specific DSP Processors", *IEEE Trans. on CAD*, pages 925–937, September 1990.

CATHEDRALL-II overview, scheduling, EXU assignment, loop folding, and a fifth-order filter for a PCM telephony application.

Goossens89

Gert Goossens, Joos Vandewalle, and Hugo De Man, "Loop Optimization in Register-Transfer Scheduling for DSP-Systems", *Proc. of the 26th DAC*, pages 826–831, June 1989.

Loop folding, and a 256-point DFT-algorithm example.

Zegers90

J. Zegers, P. Six, J. Rabaey, and H. De Man, "CGE: Automatic Generation of Controllers in the CATHEDRAL-II Silicon Compiler", *Proc. of EDAC'90*, pages 617–621, June 1989.

Controller generation.

Catthoor89

F. Catthoor, J. Van Sas, L. Inze, and H. De Man, "A Testability Strategy for Multiprocessor Architecture", *IEEE Design and Test*, pages 18–34, April 1989.

CATHEDRALL-II overview, and a testability strategy.

Rabaey88

J. Rabaey, H. De Man, J. Vanhoof, G. Goossens, and F. Catthoor, "CATHEDRAL-II: A Synthesis System for Multiprocessor DSP Systems", in *Silicon Compilation*, Daniel D. Gajski (Editor), pages 311–360, Addison-Wesley, 1988.

"Meet in the middle" design strategy, the target architecture, data path synthesis, control step scheduling and EXU assignment, data path assignment. An updated version of DeMan87, without the discussion of module generation.

Goossens87

Gert Goossens, Jan Rabaey, Joos Vandewalle and Hugo De Man, "An Efficient Microcode-Compiler for Custom DSP-Processors", *Proc. of ICCAD'87*, pages 24–27, November 1987.

Control step scheduling and register assignment.

Vanhoof87

J. Vanhoof, et. al., "A Knowledge-Based CAD System for Synthesis of Multi-Processor Digital Signal Processing Chips", *Proc. of VLSI'87*, August 1987.

DeMan87

H. DeMan, J. Rabaey, P. Six, "CATHEDRAL II: A Synthesis and Module Generation System for Multiprocessor Systems on a Chip", in *Design Systems for VLSI Circuits*, G. DeMicheli, A. Sangiovanni-Vincentelli, and P. Antognetti (Editors), pages 571–645, Martinus Nijhoff Publishers, 1987.

"Meet in the middle" design strategy, the target architecture, data path synthesis, control step scheduling, EXU assignment, data path assignment, and module generation.

DeMan86

H. De Man, J. Rabaey, P. Six, and L. Claesen, "CATHEDRAL II: A Silicon Compiler for Digital Signal Processing", *IEEE Design and Test*, pages 13–25, December 1986.

Methodology and system overview.

Six86

P. Six, L. Claesen, J. Rabaey, and H. De Man, "An Intelligent Module Generator Environment", *Proc. of the 23rd DAC*, pages 730–735, June 1986.

Module generation.

Catthoor86

F. Catthoor, J. Rabaey, G. Goossens, J. Van Meerbergen, R. Jain, H. De Man, and J. Vandewalle, "General Datapath, Controller and Inter-Communication Architectures for the Creation of a Dedicated Multi-Processor Environment", *Proc. of ISCAS'86*, pages 730–732, May 1986.

Selection of the set of EXUs in the system, and the controller architecture.

Goossens86

Gert Goossens, Jan Rabaey, Francky Catthoor, Jan Vanhoof, Rajeev Jain, Hugo De Man, and Joos Vandewalle, "A Computer-Aided Design Methodology for Mapping DSP-Algorithms Onto Custom Multiprocessor Architectures", *Proc. of ISCAS'86*, pages 924–926, May 1986.

Preliminary description of data path synthesis and control step scheduling

IMEC's CATHEDRAL-2nd System

IMEC's CATHEDRAL-2nd system is one of two successors to the CATHEDRAL-II system. It is designed for the synthesis of medium sample rate digital signal processing (DSP) chips, and it supports more flexibility in EXU composition than CATHEDRAL-II. CATHEDRAL-2nd includes transformations, data path synthesis and scheduling. See also *IMEC's CATHEDRAL-II System* and *IMEC's CATHEDRAL-III System*.

Input

Uses SILAGE.

Internal Behavioral Representation

A signal flow graph, which can represent DSP operations, multiple-precision operations, and floating-point operations.

Transformations

Expands DSP operations, multiple-precision operations, and floating-point operations into lower-level primitives.

Data Path Allocation

A CATHEDRAL-2nd-generated data path is built from a set of execution units (EXUs), as well as a set of memories (including register files and buffers), I/O units, and controller modules. The EXUs can be composed of adders, multipliers, comparators, shifters, registers, etc.

First, the designer specifies the composition and number of EXUs available. Then, during the data path synthesis phase, a chaining algorithm assigns each operation to a particular type of EXU, clustering the operations onto EXUs so as to minimize the number of clusters (control steps). However, each operation is not assigned to a specific EXU during this phase, and each variable is treated as if it were stored in a dedicated register.

Scheduling / EXU Assignment — Atomics

Uses Atomics as in CATHEDRAL-II for scheduling and EXU assignment.

Data Path Assignment

Most EXUs have two register files at their input, so EXU input values have to be clustered into those register files. A data routing algorithm is used to perform the registers assignment, first allocating all register assignments for which only one choice is available, then all other register assignments.

Examples

A pitch extractor example.

References

Verkest90

D. Verkest, L. Claesen, and H. De Man, "Correctness Proofs of Parameterized Hardware Modules in the CATHEDRAL-II Synthesis Environment", *Proc. of EDAC'90*, pages 62–66, June 1989.

System overview, transformations, data path synthesis, scheduling, and a pitch extractor example.

IMEC's CATHEDRAL-III System

IMEC's CATHEDRAL-III system is one of two successors to the CATHEDRAL-II system. It is designed for the synthesis of high-throughput chips for audio, image, and video applications, and it supports application-specific units. CATHEDRAL-III supports data path synthesis. See also *IMEC's CATHEDRAL-II System* and *IMEC's CATHEDRAL-2nd System*.

Input

Uses SILAGE.

Internal Behavioral Representation

Same signal flow graph as in CATHEDRAL-III.

Data Path Allocation

A CATHEDRAL-III-generated data path is built from a set of fast, optimized application-specific units (ASUs). ASUs perform such functions as max, min, sorting, and convolution, and are composed of standard functional units such as adders and shifters.

First, the designer specifies the composition and number of ASUs available, with the system helping to identify bottlenecks in the algorithm. Then, during the data path synthesis phase, each cluster of operations is assigned to an ASU, and individual operations are assigned to a particular functional unit within the ASU.

Scheduling / Register Allocation

Not mentioned in the literature.

Examples

A video codec example, and a Cordic processor example.

References

Note89

Stefaan Note, Francky Catthoor, Jef Van Meerbergen, and Hugo De Man, "Definition and Assignment of Complex Data-Paths Suited for High Throughput Applications", *Proc. of ICCAD'89*, pages 108–111, November 1989.

System overview, data path synthesis, and a video codec example.

Note88

Stefaan Note, Jef Van Meerbergen, Francky Catthoor, and Hugo De Man, "Automated Synthesis of a High-Speed Cordic Algorithm with the CATHEDRAL-III Compilation System", *Proc. of ISCAS'88*, pages 581–584, June 1988.

System overview, and a Cordic processor example.

Linköping University's CAMAD System

Linköping University's CAMAD (Computer Aided Modelling, Analysis, and Design of VLSI Systems) system includes scheduling, data path synthesis, and design iteration.

Input

A Pascal subset.

Internal Behavioral Representation

An extended timed Petri net (ETPN), consisting of two separate but related models: the control part, which is modelled as a timed Petri net, and the data part, which is modeled as a directed graph.

Scheduling / Data Path Synthesis / Control Allocation

The CAMAD system performs synthesis using transformations for the control part and the data part. Control part transformations can serialize or parallelize operations, and data part transformations can merge operators, connections, and constants (similar to constant folding). These transformations are controlled using a set of cost matrices, and an optimization algorithm that measures and attempts to minimize the design complexity.

Design Iteration

First, the critical path in the design is located, parallelized further using control part transformations, and transformed using data part transformations to improve its performance. Then the rest of the design is synthesized, in the order of similarity to the critical path, using data part and control part transformations to minimize the cost of the design. If a new critical path is created, this process repeats.

Examples

A small microprocessor example, an interface controller example, and a square root algorithm example.

References

Peng88a

Zebo Peng, Krzysztof Kuchcinski, and Bryan Lyles, "CAMAD: A Unified Data Path / Control Synthesis Environment", *Proc. of the IFIP Working Conference on Design Methodologies for VLSI and Computer Architecture*, pages 53–67, September, 1988.

System overview, ETPN model, design iteration, a small microprocessor example, an interface controller example, and a square root algorithm example.

Peng88b

Zebo Peng, "A Horizontal Optimization Algorithm for Data Path / Control Synthesis", *Proc. of ISCAS'88*, pages 239–242, June 1988.

Scheduling, data path synthesis, control allocation, a small microprocessor example, an interface controller example, and a square root algorithm example.

Peng86

Zebo Peng, "Synthesis of VLSI Systems with the CAMAD Design Aid", *Proc. of the 23rd DAC*, pages 278–284, June 1986.

System overview, scheduling and data path synthesis.

NTT's HARP System

NTT's HARP (Hardware Architecture Ruling Processor) system includes scheduling and data path synthesis.

Input

A subset of ANSI Fortran77, which supports constant, integer, real, logical variables and arrays, subroutines and function calls, but does not support loops with with indefinite iteration.

Internal Behavioral Representation

A dataflow graph.

Scheduling

ASAP scheduling — uses a parallelism evaluator (PE) to assign operations to control steps using a first-come first-served (FCFS) algorithm.

Data Path Synthesis

Operations are (greedily?) bound during scheduling to two types of functional units: (1) functional units performing a set of prespecified operations, and (2) functional units dynamically built during scheduling, but limited by a database of specifying permissible combinations of operations in a functional unit.

Lifetime analysis is then performed, and variables are assigned to registers using a channel-routing algorithm called the left edge algorithm to perform the binding. This is a greedy approach, assigning each value in turn to the first available register, and attempting to minimize the number of registers.

Busses are inserted whenever they would (1) have more than two inputs, (2) have more than two outputs, and (3) contain more than three interconnections to inputs and outputs.

Finally, multiplexors are inserted before any unit with multiple inputs.

Examples

Tseng's running example, and some other examples shown in tabular form.

References

Tanaka89

Toshiaki Tanaka, Tsutomu Kobayashi, and Osamu Karatsu, "HARP: Fortran to Silicon", *IEEE Trans. on CAD*, pages 649–660, June 1989.

Overview, scheduling, data path synthesis, Tseng's running example, and some other examples shown in tabular form.

Philips' PIRAMID System

Philips' PIRAMID system is an integrated environment that uses IMEC's CATHEDRAL-II system as a synthesis engine, along with a floorplanner and a module generator (grapMG). See also *IMEC's CATHEDRAL-II System*.

CATHEDRAL-II — Input, Transformations, Scheduling and Data Path Synthesis

See *IMEC's CATHEDRAL-II System* for further details.

Input

SILAGE, a Pascal-like language designed for describing DSP algorithms.

Internal Behavioral Representation

SILAGE primitives.

Transformations

Performs constant folding and common subexpression elimination.

Data Path Allocation — Jack-the-Mapper

A CATHEDRAL-II-generated data path is built from a set of execution units (EXUs) (see below), as well as a set of memories, I/O units, and controller modules. First, the designer specifies the number of EXUs of each type. Then, during the data path allocation phase, a partially-rule-based, partially-algorithmic translator assigns each operation to a particular type of EXU, assigns variables to register files, and generates a dedicated bus for each variable.

Scheduling / EXU Assignment — Atomics

Performs loop folding, overlapping the execution of successive loop iterations. Then uses list scheduling, with priority to operations on the longer critical paths. Finally, uses a vertex coloring technique to assign operations to specific EXUs.

Data Path Assignment

Values are bound to specific registers of the register file based on lifetime analysis, using a greedy algorithm. A bus-merging algorithm is then used to minimize the number of buses.

Module Generation — grapMG

Used to build the following EXUs: an ALU, an address computation unit, a multiplier-accumulator, a ROM, and a RAM. The EXUs may contain other general modules, such as register files, multiplexors, output buffers, and PLAs.

The module generator can be called by CATHEDRAL-II in a procedural way, to obtain layout, timing, area, test, and functional information. It can also be called by the floorplanner, as explained below.

Floorplanning

Accepts the structure of interconnected EXUs, calls the module generator to generate black-box views of the modules, then performs placement, global routing, and local routing. Then calls the module generator to generate detailed layout for the modules, which is used to complete the layout.

Examples

A quadrature mirror filter example.

References

Delaruelle90

A. Delaruelle, O. McArdle, J. van Meerbergen, and C. Niessen, "Synthesis of Delay Functions in DSP Compilers", *Proc. of EDAC'90*, pages 68–72, March 1990.

Synthesizing delay lines in memory.

Huisken88

J.A. Huisken, H.A.V. Janssen, P.E.R. Lippens, O. McArdle, R.H.J. Segers, P. Zegers, J.M. v. Meerbergen, "Efficient Design of Systems on Silicon with PIRAMID", in *Logic and Architecture Synthesis for Silicon Compilers (Proc. of the Int. Workshop on Logic and Arch. Synth. for Silicon Compilers)*, G. Saucier and P.M. McLellan (Editors), pages 299–311, Elsevier Science Publishers, May 1988.

PIRAMID system overview, CATHEDRAL-II, module generation, floorplanning, the EXUs designed, and a quadrature mirror filter example.

Siemens' Synthesis System

Siemens' synthesis system includes partitioning, data path synthesis and scheduling.

Input

Not specified in the literature.

Internal Behavioral Representation

Uses a dataflow graph augmented with a block structure (called a DFBS), similar to most dataflow / controlflow graphs, but with basic blocks represented explicitly.

Partitioning / Data Path Synthesis

Starts with each operation in a separate cluster, and repeatedly merges two clusters. Since each cluster corresponds to a functional unit, all operations in a cluster will be executed on that functional unit in some order. During each iteration of the algorithm, a cost is calculated for all possible merges, and the lowest-cost merge is chosen. This cost measures change in the functional unit area, register area, multiplexor and interconnect area, and in the control step schedule, based on the preliminary allocation created as the partitioning proceeds.

Once the clusters have been determined, an iterative reclustering algorithm is used to move operations between clusters, moving the operation that results in the biggest cost improvement.

Scheduling

Uses list scheduling, with expected freedom (mobility) as the priority function.

Examples

A geometric algorithm example, and a fifth-order digital elliptic wave filter example.

References

Scheichenzuber90

Josef Scheichenzuber, Werner Grass, Ulrich Lauther, and Sabine März, "Global Hardware Synthesis from Behavioral Dataflow Descriptions", *Proc. of the 27th DAC*, pages 456–461, June 1990.

Partitioning, data path synthesis, scheduling, a geometric algorithm example, and a fifth-order digital elliptic wave filter example.

Stanford's Flamel System

The Flamel system was Trickey's thesis work at Stanford University. Flamel includes transformations, scheduling, and data path synthesis.

Input

A subset of Pascal: (1) limited to a single, parameterless, non-recursive procedure, (2) supporting only single-dimensional arrays, and (3) not allowing multiplication and division except by constants that are powers of 2.

Internal Behavioral Representation

Uses a dataflow / controlflow graph, called a dacon. The initial description is represented using a block graph to show the transfers between basic blocks, and with Directed Acyclic Graphs (DAGs) to represent the data flow within each basic block. Control edges can be added to each DAG to represent execution order constraints.

Transformations

Automatically applies block-level transformations to produce the dacon with the fastest implementation. Transformations supported merge adjacent blocks, combine a block followed by two or more conditionally executed blocks into a single block, unroll loops, and move the test at the top of a loop to the bottom of the loop.

Scheduling / Data Path Synthesis

First, algebraic transformations are used to reduce the height of the dacon.

Generates an ASAP schedule with a functional unit for each operation, a register for each value, and a bus for each edge, and then attempts to fold together pairs of resources (functional units, registers, buses, etc.) that perform the same function, or that can be generalized to perform the same function. If the cost exceeds the constraint, the schedule is lengthened.

The number of resources can be constrained.

Examples

15 test programs, including bubble-sorting, string conversion, word counting, hash table management, and a finite impulse response filter.

References

Trickey87

Howard Trickey, "Flamel: A High-Level Hardware Compiler", *IEEE Trans. on CAD*, pages 259–269, March 1987.

Transformations, scheduling, data path synthesis, and the 15 test programs.

Stanford's Olympus System

Stanford University's Olympus synthesis system consists of three major components: high-level synthesis (Hercules, Hebe, and Vulcan), logic synthesis (Mercury), and module binding (Ceres). Hercules parses the behavioral description to produce the internal representation, and performs transformations; Hebe performs data path synthesis, scheduling, and design iteration. As an alternative, Vulcan performs simultaneous partitioning and control step scheduling. The system also supports an algorithmic level simulator (Ariadne) and a logic level simulator (Mercury.

Input

HardwareC, a version of C extended to support processes, interprocess communication, memory modules, parameter classes (in, out, inout), architectural registers, combinational operations, and Boolean and integer variable types. HardwareC also supports timing constraints and resource constraints.

Internal Behavioral Representations

Uses a parse tree for the transformations.

Uses a sequencing graph model, called the Sequencing Intermediate Form (SIF), for scheduling and allocation. The vertices of the graph represent operations to be performed, and the edges represent predecessor / successor relationships (which are subject to the data dependencies between operations). Vertices may either be state vertices, requiring at least one cycle for execution, or stateless, requiring no time to execute.

Transformations

User-driven transformations allow the user to expand procedure calls inline, or to specify the library modules that should be used to implement particular operations.

Automatic transformations include constant and variable folding, common subexpression elimination, dead code elimination, reduction of constant conditionals, and loop unrolling.

Partitioning

Constraints can be placed on the area-cost and pinout-cost of each block, and on the overall latency. Two partitioning algorithms are supported: one using simulated annealing, and one based on the Kernighan-Lin algorithm. The cost functions for these algorithms consider the communication costs, latency, area, and pinout.

Data Path Synthesis

Operations are mapped onto Boolean equations, and then onto combinational library modules. Variables are mapped onto registers only if they are declared as architectural registers, or if they are used in a loop; otherwise, they are implemented as wires.

The design space is explored using either an exact or a heuristic search of the assignments, using cost criteria that include area, interconnection, and delay.

Scheduling

Ku90a, Ku90b

Uses an iterative graph-based technique scheduling technique that supports timing constraints, and operations with unbounded delays.

DeMicheli88

A ripple-through-control model is used to produce a schedule with a minimal number of states and transitions between states.

Design Iteration

First, data path synthesis and scheduling are performed assuming zero delay and area for each combinational operation. Then, after logic synthesis has been performed, those delay and area values can be fed back to guide the next iteration of the system.

Examples

The Intel 8251, the FRISC microprocessor, the MCS6502, and a fifth-order digital elliptic wave filter example.

References

Gupta90

Rajesh Gupta and Giovanni De Micheli, "Partitioning of Functional Models of Synchronous Digital Systems", *Proc. of ICCAD'90*, pages 216–219, November 1990.

Partitioning, and a fifth-order digital elliptic wave filter example.

DeMicheli90

Giovanni De Micheli, David C. Ku, Frédéric Mailhot, and Thomas Truong, "The Olympus Synthesis System", *IEEE Design and Test*, pages 37–53, October 1990.

Olympus overview, HardwareC, transformation, data path synthesis, scheduling, logic synthesis, and module binding.

Ku90a

David Ku and Giovanni De Micheli, "Relative Scheduling under Timing Constraints", *Proc. of the 27th DAC*, pages 59–64, June 1990.

Scheduling with timing constraints.

Ku90b

David Ku and Giovanni De Micheli, "High-level Synthesis and Optimization Strategies in Hercules and Hebe", *Proc. of EuroASIC'90*, pages 124–129, May 1990.

Hercules/Hebe overview, HardwareC, transformations, the SIF, data path synthesis, and scheduling.

Ku90c

David Ku and Giovanni De Micheli, *HardwareC — A Language for Hardware Design, Version 2.0*, Computer Systems Laboratory Technical Report CSL-TR-90-419, April 1990.

A complete description of HardwareC.

DeMicheli88

Giovanni De Micheli and David C. Ku, "HERCULES — A System for High-Level Synthesis", *Proc. of the 25th DAC,* pages 483–488, June 1988.

Hercules overview, transformations, scheduling, data path synthesis, the Intel 8251, the FRISC microprocessor, and the MCS6502.

Tsing Hua University's ALPS / LYRA / ARYL System

Tsing Hua University's synthesis system includes scheduling (ALPS) and data path synthesis (LYRA and ARYL).

Input

Not mentioned in the literature.

Internal Behavioral Representation

A data flow graph.

Scheduling / Functional Unit Allocation — ALPS

Hwang90a

First, list scheduling is used to determine the upper bound on the number of control steps, and ASAP and ALAP scheduling are used to determine the minimum and maximum start times for each operation. Then integer linear programming is used to schedule the operations into control steps, and allocate functional units, subject to resource constraints, and minimizing the number of control steps. To make this integer linear programming problem more manageable, the schedule is broken equally into a fixed number of zones, and each zone is scheduled sequentially.

Lee89

First, list scheduling and ASAP and ALAP scheduling are used as described above. Then integer linear programming is used to schedule the operations into control steps, and allocate functional units, while minimizing the functional unit cost.

Extensions to the basic algorithm handle pipelined functional units, a pipelined data path, and loop folding.

Data Path Synthesis — LYRA & ARYL

LYRA does register allocation, functional unit binding, connection allocation, in that order, while ARYL does functional unit binding, register allocation, and connection allocation.

Register allocation: LYRA first determines the minimum number of registers necessary, using lifetime analysis to find the maximum number of lifetimes crossing any one control step. Then it builds a graph for a set of mutually unsharable variables, indicating which variables which can not share a register due to overlapping lifetimes, and uses a weighted matching method to bind the variables to registers.

Functional unit binding: LYRA first builds a table indicating the cost gain of combining all possible pairs of operations into one functional unit, where the gain is based on the need for additional multiplexors. Then it builds a graph representing possible bindings of operations to functional units, and uses a weighted matching method to bind the operations to functional units one control step at a time.

Connection allocation: Iteratively constructs multiplexor trees as necessary.

Examples

MAHA's example, a fifth-order digital elliptic wave filter example, and a second-order differential equation example.

References

Hwang90a

Cheng-Tsung Hwang, Yu-Chin Hsu, and Youn-Long Lin, "Optimum and Heuristic Data Path Scheduling Under Resource Constraints", *Proc. of the 27th DAC*, pages 65–70, June 1990.

Scheduling and functional unit allocation, MAHA's example, a fifth-order digital elliptic wave filter example.

Hwang90b

Chu-Yi Huang, Yen-Shen Chen, Youn-Long Lin, and Yu-Chin Hsu, "Data Path Allocation Based on Bipartite Weighted Matching", *Proc. of the 27th DAC*, pages 499–504, June 1990.

Data path synthesis, a second-order differential equation example, and a fifth-order digital elliptic wave filter example.

Lee89

Jiahn-Hung Lee, Yu-Chin Hsu, and Youn-Long Lin, "A New Integer Linear Programming Formulation for the Scheduling Problem in Data Path Synthesis", *Proc. of ICCAD'89*, pages 20–23, November 1989.

Scheduling and functional unit allocation, MAHA's example, and a fifth-order digital elliptic wave filter example.

Univ. of California at Berkeley's HYPER System

The University of California at Berkeley's HYPER system is aimed at synthesizing real-time applications, and includes transformations, scheduling, data path synthesis, module binding, and controller design.

Input

An extended version of SILAGE, a signal-flow oriented language designed for describing DSP algorithms.

Internal Behavioral Representation

A dataflow / controlflow graph.

Transformations

Applies optimizing compiler transformations, including constant folding, common subexpression elimination, dead code elimination, loop retiming, loop pipelining, loop unrolling, and loop jamming.

Scheduling / Data Path Synthesis

First, the upper and lower bounds on the number of functional units, registers, and buses is calculated, to determine the initial number of resources available. Then each control step is scheduled in turn. For each control step, a local ratio is calculated (for the current control step, the ratio of the number of available resources of each type over the number of required resources of each type), a global ratio is calculated (similar, but for all operations still unscheduled). The algorithm then schedules and binds the operation which keeps the most critical ratio (the smallest one) as high as possible, continuing until that control step is completed.

Module Binding

Uses a set of small transformations, including multiplexor reduction, allocation of assignment operations, and data path partitioning, plus several translation steps, which perform the module binding.

Controller Design

Applies transformations to the control step schedule to remove dummy states, and allocates control registers, interface logic, and a finite state machine to build the controller. Optimizations are then applied to reduce the size of the finite state machine, and to simplify the wiring between the controller and the data path.

Examples

A fifth-order WDF filter example, and a Hidden Markov Model based speech recognition system example.

References

Chu89

Chi-Min Chu, Miodrag Potkonjak, Markus Thaler, and Jan Rabaey, "HYPER: An Interactive Synthesis Environment for High Performance Real Time Applications", *Proc. of ICCD'89*, pages 432–435, October 1989.

System overview, module binding, and controller design, and a Hidden Markov Model based speech recognition system example.

Potkonjak89

Miodrag Potkonjak and Jan Rabaey, "A Scheduling and Resource Allocation Algorithm for Hierarchical Signal Flow Graphs", *Proc. of the 26th DAC*, pages 7–12, June 1989.

Scheduling, data path synthesis, and a fifth-order WDF filter example.

Univ. of California at Berkeley's Synthesis System

The University of California at Berkeley's synthesis system includes transformations, scheduling and data path synthesis.

Input

The C language.

Internal Behavioral Representation

A serial / parallel / disjoint code sequence, represented textually in a Lisp-based syntax.

Transformations

Uses various compiler transformations, including constant folding and dead code elimination.

Scheduling / Data Path Synthesis

Views the scheduling / data path synthesis problem as one of placing operations onto a two-dimensional grid, where columns in the grid represent functional units, and rows in the grid represent control steps. Uses a simulated annealing approach to interchange operations in the grid, move operations from one location in the grid to another, and switch the order of input to commutative operations. Extensions to the basic algorithm handle disjoint code sequences, loops, loop unwinding, and pipeline synthesis.

Examples

Tseng's running example, a MOSFET model evaluation routine, a code sequence example, and a fifth-order digital elliptic wave filter example.

References

Devadas89

Srinivas Devadas, "Algorithms for Hardware Allocation in Data Path Synthesis", *IEEE Trans. on CAD*, pages 768–781, July 1989.

Scheduling, data path synthesis, extensions to the basic algorithms, and a fifth-order digital elliptic wave filter example.

Devadas87a

Srinivas Devadas and A. Richard Newton, "Data Path Synthesis from Behavioral Descriptions: An Algorithmic Approach", *Proc. of ISCAS'87*, pages 398–401, May 1987.

Scheduling, pipelining a non-pipelined design by partitioning the control step schedule into phases while adding the minimum amount of new hardware, and a code sequence example.

Devadas87b

Srinivas Devadas and A. Richard Newton, "Algorithms for Hardware Allocation in Data Path Synthesis", *Proc. of ICCD'87*, pages 526–531, October 1987.

Scheduling, data path synthesis, Tseng's running example, and a MOSFET model evaluation routine.

Univ. of California at Irvine's VSS

The University of California at Irvine's VHDL Synthesis System (VSS) produces Register Transfer level designs, which can then passed on to the Microarchitecture and Logic Optimizer (MILO) system for optimization and library binding. The VSS includes transformations, scheduling, data path synthesis, and functional synthesis.

Input

VHDL behavioral and dataflow descriptions at the logic, Register Transfer, and algorithmic levels. Signals can be typed, can have a specified bit width, and can have clocking and sensitivity information specified.

Internal Behavioral Representation

Uses a hierarchical "control/data flowgraph", with nodes representing READs, WRITEs, and operators, and nets representing connectivity and signal attributes. A graphical display allows the user to view this flowgraph.

Transformations

Uses "cleanup rules" to eliminate redundant operations in the flowgraph, and "optimization rules" to replace behavioral constructs with others more closely matching library components. Also uses transformation for loop unwinding and loop pipelining.

Scheduling

First, uses percolation scheduling to allow data-independent operations to "percolate" toward earlier control steps, not considering resource constraints. To meet resource constraints, the mobility of each operation in the control step is computed, and operations with higher mobility are delayed until later control steps.

Data Path Synthesis

Minimizes the number of functional units, registers, and connections by sharing.

In Register Transfer mode, minimizes the number of functional units and connections by sharing.

Each node is then replaced by a one or more components from a generic component library.

Functional Synthesis

Functional synthesis is the synthesis of concurrent functional components, such as ALUs, from a behavioral description. Since these components are not state machines, control step scheduling is not applicable. Since all variables must be stored in registers, register merging is not applicable. Since there is only a single "state", functional units are not shared between states. However, functional units can be shared between conditional branches, and expressions can be reduced.

The Component Synthesis Algorithm (CSA) parses the behavioral description into a dataflow / controlflow graph, and merges expressions to match library components using a clique partitioning algorithm guided by a branch-and-bound control strategy.

Examples

A full adder example, the Intel I8212, a bus interface example, a controlled counter example, a fifth-order digital elliptic wave filter example, a 16-point FIR filter example, a 256-point DFT-algorithm example, and a B-spline FIR filter example.

References

Rundensteiner90

Elke A. Rundensteiner, Daniel D. Gajski, and Lubomir Bic, "The Component Synthesis Algorithm: Technology Mapping for Register Transfer Descriptions", *Proc. of ICCAD'90*, pages 208–211, November 1990.

Potasman90

Roni Potasman, Joseph Lis, Alexandru Nicolau, and Daniel Gajski, "Percolation Based Synthesis", *Proc. of the 27th DAC*, pages 444–449, June 1990.

Loop transformations, scheduling, a fifth-order digital elliptic wave filter example, a 16-point FIR filter example, a 256-point DFT-algorithm example, and a B-spline FIR filter example.

Lis89

Joseph S. Lis and Daniel D. Gajski, "VHDL Synthesis Using Structured Modeling", *Proc. of the 26th DAC*, pages 606–609, June 1989.

Modeling designs at different levels of abstraction, and synthesis from those different design models.

Lis88

Joseph S. Lis and Daniel D. Gajski, "Synthesis from VHDL", *Proc. of ICCD'88*, pages 378–381, October 1988.

Use of VHDL for synthesis, system overview, data path synthesis (brief), a full adder example, the Intel I8212, a bus interface example, and a controlled counter example.

Orailoglu86

Alex Orailoglu and Daniel D. Gajski, "Flow Graph Representation", *Proc. of the 23rd DAC*, pages 503–509, June 1986.

A description of an older version of the dataflow graph used by the VSS, with provisions for more accurate modeling of memory accesses.

Univ. of Illinois' Chippe / Slicer / Splicer System

The University of Illinois' synthesis system consists of three parts: Chippe, Slicer, and Splicer. Chippe controls the synthesis process, Slicer is the scheduler, and Splicer is the data path allocator. Chippe was Brewer's thesis work, and Slicer / Splicer was Pangrle's thesis work, both at the University of Illinois at Urbana-Champaign.

Brewer and Pangrle have since continued to work on the system at the University of California at Santa Barbara and at Pennsylvania State University, respectively.

Input

Uses a version of Pascal, extended to support type declarations (including bit widths), register declarations, and port declarations.

Internal Behavioral Representation

Uses a dataflow / controlflow graph based on an Algorithmic State Machine (ASM) chart. Each straight-line block of code is represented as a dataflow graph, a list of input variables, a list of output variables, and a list of successor blocks.

Design Iteration — Chippe

During the synthesis process, the Chippe expert system controls the synthesis tools using resource constraints ("knobs"). The resulting design is evaluated by a set of quality measures ("guages"), and compared to the design goal; tradeoffs can be made by adjusting the knobs for another iteration of the synthesis tools. Measures of the quality of the design include estimates of area, power dissipation, and execution time. Other measures include "overlap", the number of states for which two units are active in parallel, "dead time", the amount of the system clock cycle that is unused, and "bus usage". The tradeoffs include function unit selection, serial / parallel scheduling tradeoffs, and system clock speed.

Schalloc / Splicer Synthesis Path

Scheduling / Data Path Synthesis — Schalloc / Splicer

The Schalloc algorithm maintains two sets of operations: a ready-list, which is the list of operations that are data ready, and a max-micro-group, which is a set of operations that can be executed in one control step. Initially, the max-micro-group is empty, and the ready-list contains all operations that are data ready for the first control step. A branch-and-bound, recursive algorithm is then used for scheduling: at each step in the algorithm, all possible max-micro-group and ready-list pairs are determined, Splicer is used to perform the data path synthesis for each, the best solution is retained, and the process repeats until the ready-list is exhausted or the cost exceeds the cost of the best solution.

Slicer / Splicer Synthesis Path

Scheduling — Slicer

List scheduling, with operations ranked according to their mobility, and operations with lower mobility scheduled before scheduling operations with higher mobility. An operation's mobility is the number of control steps that it can possibly be delayed from its ASAP control step, i.e. the difference between its ALAP control step and its ASAP control step. Thus operations with a mobility equal to zero are on the critical path, and are scheduled first. If two nodes have the same mobility, the node with the most successors is scheduled first.

The number of functional units and the operations that they can perform can be limited. If a resource limit is encountered, extra control steps are added.

Data Path Synthesis — Splicer

Uses point-to-point connectivity, buses, and two-level multiplexors, with the user able to specify bus-style or multiplexor-style. For buses, assumes that a register connects to an input bus, the input bus connects to a functional unit, the functional unit connects to an output bus, and the output bus connects to a register.

Uses a backtracking heuristic based on branch and bound with a depth-first search. Cost functions are based on the four levels of interconnection below.

The user can also specify the lookahead (the number of states beyond the present state to be considered simultaneously for connectivity binding).

Register to Input Bus Connections

Connects all registers that are used for input to input buses, first trying buses that are already connected to the registers before adding new input buses.

Input Bus to Function Unit Connections

Connects all input buses to function unit inputs, first trying to use any existing function units that perform the needed function.

Function Unit to Output Bus Connections

Connects all function unit outputs to output buses, first trying buses that are already connected to function units before adding new output buses.

Output Busses to Registers

Connects output buses to registers, dynamically performing register allocation.

Examples

Chippe

A differential equation example, a fifth-order digital elliptic wave filter example, and a TMS320 example.

Schalloc / Splicer

A differential equation example, a fifth-order digital elliptic wave filter example, and a temperature controller example.

Slicer / Splicer

Tseng's running example, a differential equation example, a wave filter example, and the PDP-8.

References

Brewer90

Forrest Brewer and Daniel Gajski, "Chippe: A System for Constraint Driven Behavioral Synthesis", *IEEE Trans. on CAD*, pages 681–695, July 1990.

Chippe overview, design tradeoffs, a differential equation example, a fifth-order digital elliptic wave filter example, and a TMS320 example.

Berry90

Neerav Berry and Barry M. Pangrle, "Schalloc: An Algorithm for Simultaneous Scheduling & Connectivity Binding in a Datapath Synthesis System", *Proc. of EDAC'90*, pages 78–82, March 1990.

Scheduling (Schalloc), a differential equation example, a fifth-order digital elliptic wave filter example, and a temperature controller example.

Pangrle88

Barry M. Pangrle, "Splicer: A Heuristic Approach to Connectivity Binding", *Proc. of the 25th DAC*, pages 536–541, June 1988.

Data path synthesis, Tseng's running example, a differential equation example, and a wave filter example.

Pangrle87a

Barry Michael Pangrle and Daniel D. Gajski, "Design Tools for Intelligent Silicon Compilation", *IEEE Trans. on CAD*, pages 1098–1112, November 1987.

Scheduling (Slicer), data path synthesis, Tseng's running example, a differential equation example, and the PDP-8. Essentially Pangrle87b + Pangrle88.

Pangrle87b

Barry Michael Pangrle and Daniel D. Gajski, "Slicer: A State Synthesizer for Intelligent Silicon Compilation", *Proc. of ICCD'87*, pages 42–45, October 1987.

Scheduling (Slicer) and a differential equation example.

Brewer87

Forrest D. Brewer and Daniel D. Gajski, "Knowledge Based Control in Micro-Architecture Design", *Proc. of the 24th DAC*, pages 203–209, June 1987.

Organization of the system and a differential equation example.

Pangrle87c

Barry Michael Pangrle, *A Behavioral Compiler for Intelligent Silicon Compilation*, PhD Thesis, Dept. of Computer Science, University of Illinois, 1987.

Other synthesis systems, scheduling (Slicer), data path synthesis, Tseng's running example, a differential equation example, and the PDP-8.

Gajski87

Daniel D. Gajski and Forrest D. Brewer, "Towards Intelligent Silicon Compilation", in *Design Systems for VLSI Circuits*, G. DeMicheli, A. Sangiovanni-Vincentelli, and P. Antognetti (Editors), pages 365–383, Martinus Nijhoff Publishers, 1987.

A model for design and design iteration, and a shift-add multiplier example. Similar to Brewer86.

Pangrle86

Barry M. Pangrle and Daniel D. Gajski, "State Synthesis and Connectivity Binding for Microarchitecture Compilation", *Proc. of ICCAD'86*, pages 210–213, November 1986.

System overview, Tseng's running example, and a differential equation example.

Brewer86

Forrest D. Brewer and Daniel D. Gajski, "An Expert-System Paradigm for Design", *Proc. of the 23rd DAC*, pages 62–68, June 1986.

A model for design and design iteration, and a shift-add multiplier example.

Univ. of Illinois' IBA System

The Univ. of Illinois' IBA (Interleaved Binder and Allocator) system consists of several parts: Illinois Mixed Behavior / Structure Language (IMBSL), which performs data path synthesis; RLEXT (Register Level Exploration Tool), which allows a user to manually modify a Register-Transfer level design, and then automatically repairs any errors or omissions so that the final result matches the specified behavior; COD, a control unit synthesizer; LE (Layout Estimator); and Fasolt, a Register-Transfer level datapath optimizer. See also *Univ. of Southern California's ADAM System* — Knapp used to be involved with that system.

Input

A behavioral specification and a partial structure specification, both written in IMBSL (Illinois Mixed Behavior / Structure Language).

Internal Behavioral Representation

A modified version of the Univ. of Southern California's Design Data Structure (DDS).

Scheduling

None — left to another tool, or to the user.

Data Path Synthesis — IMBSL

First, functional units are allocated, using cost models to produce the minimum cost implementation. Then interconnections are added iteratively: in each iteration, unbound values are examined, and if they can not be routed through existing interconnections, they "vote" for changes that will produce an acceptable routing. At the end of the pass, the interconnection changes with the most "votes" are implemented, and another pass is made.

Interactive Design Transformation — RLEXT

The user can modify a data path by adding and/or deleting functional units and interconnection. At any time, the user can invoke the procedure fixit, which repairs the operator binding and interconnection, so that the original behavior is preserved.

The user can also modify the control step schedule, shifting operations, inserting or deleting control steps, and compressing sequences of control steps.

Automatic Design Transformation — Fasolt

First, Fasolt uses LE to generate a macrocell-oriented floorplan and global routine. Fasolt then merges pairs of wires (buses) using a cost gain / performance penalty analysis, modifies the control step schedule as necessary, and uses RLEXT to clean up any resulting inconsistencies in the design. This process is then repeated as long as Fasolt succeeds in reducing the area estimate determined by LE.

Examples

A small microprocessor, and the MCS6502.

References

Binger90

David Binger and David W. Knapp, "Automatic Synthesis of a Dual-PLA Controller with a Counter", *Proc. of Micro-23*, pages 149–157, November 1990.

Constructing two-PLA finite state machine controllers, and the MCS6502.

Knapp90a

David W. Knapp, "Feedback-Driven Datapath Optimization in Fasolt", *Proc. of ICCAD'90*, pages 300–303, November 1990.

Automatic design transformation using Fasolt and LE.

Knapp90b

David W. Knapp and Marianne Winslett, "A Formalization of Correctness for Linked Representations of Datapath Hardware", in *Formal VLSI Specification and Synthesis*, L.J.M. Claesen (Editor), pages 3–22, North-Holland, 1990.

A formal representation for Register-Transfer level designs, including the design library, the dataflow behavior, the timing model, the structural model, and the bindings between these representations; and RLEXT overview.

Knapp89a

David W. Knapp, "Manual Rescheduling and Incremental Repair of Register-Level Datapaths", *Proc. of ICCAD'89*, pages 58–61, November 1989.

Using RLEXT to incrementally repair the control step schedule after it has been manually modified by the user.

Knapp89b

David W. Knapp, "An Interactive Tool for Register-Level Structure Optimization", *Proc. of the 26th DAC*, pages 598–601, June 1989.

Univ. of Illinois' version of the DDS, RLEXT design transformation, and a small microprocessor.

Winslett89

Marianne Winslett, David Knapp, Keith Hall, and Gio Wiederhold, "Use of Change Coordination in an Information-Rich Design Environment", *Proc. of the 26th DAC*, pages 252–257, June 1989.

Representing a design using the Univ. of Illinois' version of the DDS, and coordinating changes between the behavioral, timing, structural, and physical domains.

Knapp89c

David W. Knapp, "Synthesis from Partial Structure", in *Design Methodologies for VLSI and Computer Architecture*, D. A. Edwards (Editor), pages 35–51, North-Holland, 1989.

Univ. of Illinois' version of the DDS, IMBSL data path synthesis, and the MCS6502.

Univ. of Karlsruhe's DSL Synthesis System / CADDY System

The University of Karlsruhe's DSL Synthesis System (also called the Carlsruhe Digital Design System (CADDY)) includes transformations, data path synthesis, and scheduling. See also *IBM's Yorktown Silicon Compiler* and *IBM's HIS System* — Camposano later became involved with those systems.

Input

DSL (Digital Systems Specification Language), a Pascal-like language. This language supports both imperative (sequential, procedural) descriptions and applicative (functional, dataflow) descriptions. A DSL program may include many imperative parts, but only one applicative part, which may be used to specify global actions such as resets, interrupts, etc. that may occur at any point in time. DSL supports process-level concurrency, and communication through common variables. Both global and local timing constraints can be described.

Internal Behavioral Representation

Uses three graphs that share the same vertices, with each vertex representing either an operation or variable in the DSL program. A controlflow graph represents the predecessor-successor relationships between the vertices, a dataflow graph represents the data dependencies between the vertices, and a third graph represents timing constraints between the vertices.

From this internal representation, a dataflow graph is constructed for the imperative parts of the input description, with new variables inserted to distinguish different values held by the same variable.

Transformations

Uses transformations to eliminate useless registers, replace addition by 1 with an increment operation, unroll loops, and perform dead code elimination, etc.

Scheduling / Data Path Synthesis Path — Krämer90

Scheduling / Functional Unit Allocation

Uses list scheduling with resource constraints, trying to minimized the total delay though the design. At each iteration of the list scheduling algorithm, all possible allocations of operations to functional units is considered. The unscheduled operations are scheduled using ASAP and ALAP scheduling without resource constraints to determine their freedom (mobility), and this schedule, along with a list of the available functional units, is used to determine the expected delay. This expected delay is then used by the list scheduling algorithm to schedule the operation with the minimal cost.

Data Path Synthesis

First, the user manually partitions the design onto different processors or pipeline stages.

Second, an interconnection network, consisting of global buses and multiport memories, is constructed to allow the processors to communicate. The buses are allocated using a graph coloring algorithm.

Finally, the data paths for the individual processors are constructed, using the functional units, registers, and buses. First, variables are assigned to registers using lifetime analysis and graph coloring. Then the interconnect costs are minimized using a graph coloring algorithm that takes into account preferred interconnect styles. Finally, operations are assigned to specific functional units, and data transfers are assigned to buses, again using graph coloring algorithms and preferences.

Data Path Synthesis / Scheduling Path — 1989 and Before

Data Path Synthesis (Imperative Description)

First, operations are mapped directly onto individual functional units, although multiple operations can be explicitly specified to be executed by the same functional unit. Then global and local optimizations are performed. These include folding multiple operations onto a single functional unit, and tradeoffs between specialized functional units (e.g., an adder or subtracter) and more general functional units (e.g., an ALU).

Registers are added at loop boundaries, and as necessary due to the folding of operations into functional units. Local transformations can then be used to reduce the number of registers.

Some large multiplexors are replaced by tri-state buses.

The resulting structure is described in STRUDEL (Structure Description Language).

Scheduling (Imperative Description)

States are created in the schedule based on the control structure, on the data dependencies between the operations, when registers are written, and when interface pins are accessed. A rule-based system is then used to merge states in the control schedule.

Examples

Scheduling / Data Path Synthesis Path

A fifth-order digital elliptic wave filter example, a Kalman filter example, and a towers of Hanoi example.

Data Path Synthesis / Scheduling Path

A decoder example, a priority encoder example, the MC68000, a multiplier example, and a cell for a sorting stack example.

References

Krämer90

Heinrich Krämer and Wolfgang Rosenstiel, "System Synthesis using Behavioural Descriptions", *Proc. of EDAC'90*, pages 277–282, March 1990.

Scheduling, data path synthesis, a fifth-order digital elliptic wave filter example, a Kalman filter example, and a towers of Hanoi example.

Camposano89

Raul Camposano and Wolfgang Rosenstiel, "Synthesizing Circuits from Behavioral Descriptions", *IEEE Trans. on CAD*, pages 171–180, February 1989.

The DSL language, mapping onto a data flow graph, data path synthesis, control step scheduling, a decoder example, a priority encoder example, and the MC68000.

Camposano86

Raul Camposano and Arno Kunzmann, "Considering Timing Constraints in Synthesis from a Behavioral Description", *Proc. of ICCD'86*, pages 6–9, October 1986.

Specifying and evaluating timing constraints.

Rosenstiel86

Wolfgang Rosenstiel, "Optimizations in High Level Synthesis", *Microprocessing and Microprogramming (18)*, pp. 543-549, North Holland, 1986.

Transformations, logic minimization, and Register-Transfer level hardware transformations to improve speed and area consumption.

Camposano85a

Raul Camposano and Wolfgang Rosenstiel, "A Design Environment for the Synthesis of Integrated Circuits", *Proc. of the 11th EUROMICRO Symp. on Microprocessing and Microprogramming*, pages 211–215, September 1985.

Overview of transformations, allocation, scheduling, module generation, and layout synthesis.

Rosenstiel85

W. Rosenstiel and R. Camposano, "Synthesizing Circuits form Behavioral Level Specifications", in *Computer Hardware Description Languages and their Applications (Proc. of the 7th Int.*

Conf. on CHDLs), C.J. Koomen and T. Moto-oka (Editors), pages 391–402, North Holland, August 1985.

System overview and data path synthesis.

Camposano85b

Raul Camposano, "Synthesis Techniques for Digital Systems Design", *Proc. of the 22nd DAC*, pages 475–481, June 1985.

Design representation and optimizations.

Camposano85c

Raul Camposano, Arno Kunzmann, and Wolfgang Rosenstiel, "Automatic Data Path Synthesis from Behavioural Level Descriptions in DSL", in *VLSI: Algorithms and Architectures (Proc. of the Int. Workshop on Parallel Computing and VLSI, May, 1984)*, P. Bertolazzi and F. Luccio (Editors), pages 233–242, North Holland, 1985.

Overview of prototype system, a multiplier example and a cell for a sorting stack example.

Camposano84

Raul Camposano, Arno Kunzmann, and Wolfgang Rosenstiel, "Automatic Data Path Synthesis from DSL Specifications", *Proc. of ICCD'84*, pages 630–635, October 1984.

Overview of prototype system and a cell for a sorting stack example.

Univ. of Kiel's MIMOLA System

MIMOLA (Machine Independent Microprogramming Language) refers to both a design language and a design method. MSS1 was Zimmermann's original software system; MSS2 is the current version at the University of Kiel. Honeywell also has a new version. The MIMOLA system includes scheduling, data path synthesis, and controller design. See also *IIT Delhi's Synthesis System* — Marwedel is also involved with that system — and *Honeywell's V-Synth System*, which adds a VHDL front end to MSS1.

Input

Uses the MIMOLA language, a Pascal-like language that includes recursive procedure calls and multi-dimensional arrays.

Internal Behavioral Representation

Uses an intermediate language "similar to intermediate tree languages used in compiler development projects" [Marwedel85].

Scheduling

Statements are decomposed into simple statements, then processed by basic block using a variation on Dasgupta's 1976 pairwise comparison algorithm.

Resource constraints can be placed on the number of available ALUs and the number of memory ports.

Data Path Synthesis

Resister Binding (called register assignment)

Scans each control step, allocating registers in a greedy fashion.

Functional Unit Binding (called module selection)

Three types of ALUs can be generated: (1) ALUs completely specified in the initial description, (2) new ALUs based on the operations needed, and (3) ALUs of certain prespecified types. The present MIMOLA system concentrates on the latter.

Uses an integer programming algorithm to select functional units that minimize the overall functional unit cost. Note that individual operations are not yet bound to specific functional units.

Functional Unit Binding / Interconnect Binding

Uses a branch-and-bound algorithm to bind operations to functional units so as to minimize the total number of paths. The control steps are bound starting with the most complex instruction.

Controller Design

Generates a microprogrammed controller with horizontal microcode, although some resources are allowed to share microword bits. Microwords with long fields and fields with conflicts are scheduled first.

Examples

A scientific subroutine package example, a Siemens 7.000-type machine, a processor for graphic layout, and an operating system kernel example.

References (1983 and Later)

Marwedel90

P. Marwedel, "Matching System and Component Behaviour in MIMOLA Synthesis Tools", *Proc. of EDAC'90*, pages 146–156, March 1990.

System overview.

Marwedel86

Peter Marwedel, "A New Synthesis Algorithm for the MIMOLA Software System", *Proc. of the 23rd DAC*, pages 271–277, June 1986.

Scheduling, data path synthesis, and controller design.

Krüger86

Gerd Krüger, "Automatic Generation of Self-Test Programs — A New Feature of the MIMOLA Design System", *Proc. of the 23rd DAC*, pages 378–384, June 1986.

Generating self-test programs, and a simple processor example.

Marwedel85

Peter Marwedel, "The MIMOLA Design System: A Design System Which Spans Several Levels", in *Methodologies for Computer System Design*, W.K. Giloi and B.D. Shriver (Editors), pages 223–237, North Holland, 1985.

System overview, a scientific subroutine package example, a Siemens 7.000-type machine, a processor for graphic layout, and an operating system kernel example.

Marwedel84a

Peter Marwedel, "A Retargetable Compiler for a High-Level Microprogramming Language", *Proc. of Micro-17*, pages 267–274, October 1984.

System overview, the MIMOLA language, and older synthesis algorithms.

Marwedel84b

Peter Marwedel, "The MIMOLA Design System: Tools for the Design of Digital Processors", *Proc. of the 21st DAC*, pages 587–593, June 1984.

System overview, a scientific subroutine package example, an IBM 370, and an operating system kernel example.

Univ. of Paderborn's PARBUS System

The University of Paderborn's PARBUS system includes scheduling, partitioning, and data path synthesis. The system is aimed at producing designs with two partitioned buses.

Input

Not described in the literature.

Internal Behavioral Representation

Not described in the literature.

Scheduling

List scheduling, with constraints on the number of functional units, chaining, multi-cycle operations, and support for pipelined functional units.

Partitioning

Clusters operations into partitions using a graph coloring algorithm, maximizing the number of common inputs and outputs within each partition. Symbolic registers are then added to store operand and results.

The partitions are then arranged in a linear order. A communications graph is constructed, with nodes representing the partitions, and weights on the edges representing the amount of communication. A branch-and-bound algorithm is then used to order the partitions so as to create a Hamiltonian path with the maximum weight.

Data Path Synthesis

The registers in each partition are grouped, and the ports of the registers and functional units are assigned to the buses using a communications graph and a maximum matching algorithm. Then inter-partition communications are determined, a communication interference graph is built and used to satisfy the global communications between partitions, and the symbolic registers are mapped onto physical registers.

Examples

A differential equation example, and a fifth-order digital elliptic wave filter example.

References

Ewering90

Christian Ewering, "Automatic High Level Synthesis of Partitioned Busses", *Proc. of ICCAD'90*, pages 304–307, November 1990.

Scheduling, partitioning, data path synthesis, a differential equation example, and a fifth-order digital elliptic wave filter example.

Univ. of Southern California's ADAM System

The University of Southern California's Advanced Design Automation (ADAM) system includes scheduling, data path synthesis, and controller specification generation. Two general-purpose synthesis paths are supported: MAHA/REAL, which performs scheduling and register allocation, and MAHA/MABAL, which performs scheduling and data path synthesis. ADAM also includes programs for planning, area estimation, and module selection. A related project, SEHWA, synthesizes pipelined designs. See also *Univ. of Illinois' IMBSL / RLEXT System* — Knapp is now involved with that system.

Input

The PHRAN-SPAN natural language interface for system-level specifications, the AGIS graphics interface for directly manipulating the Design Data Structure, the 3DIS data base interface, and the SLIDE hardware description language.

Internal Behavioral Representation

Uses the Design Data Structure (DDS) to represent behavioral, structural, and physical information.

The data flow model represents operations, operation inputs, and data dependencies between operations using a bipartite directed acyclic graph.

The control and timing model represents events and relations between events, also using a directed acyclic graph.

The DDS also has the capability to represent timing constraints, and both synchronous and asynchronous behavior.

Planning and Prediction

Based on the design's specifications and constraints, a plan, or abstract sequence of tasks, is constructed to guide the design process. During the construction of this plan, estimates (called "predictors") of the area, speed, etc. of the final design are used to explore the design space. When the plan is completed, the actual tools are used to construct the final design.

Scheduling / Functional Unit Binding — MAHA

First, Park's Clocking Scheme Synthesis Package (CSSP) is used to find the operations on the critical path and the optimal number of steps in the critical path, and functional units are allocated and bound to those operations. Then, the freedom (mobility) of each operation is computed (the difference between the time when the input values are needed and the time when the results are needed, less the delay for the operation), and the operations with the smallest freedom are scheduled first. If necessary, additional functional units are added subject to the cost constraints and the freedoms are recalculated. If additional functional units can not be added, additional time steps are added to the critical path (subject to the time constraints) and the allocation begins again.

Register Allocation — REAL

Treats values as wires and registers as tracks, and uses a channel-routing algorithm called the left edge algorithm to perform the binding; this algorithm has been proven optimal. This is a greedy approach, assigning each value in turn to the first available register, and attempting to minimize the number of registers. Extensions to the algorithm support conditional branches and pipelined data paths.

Data Path Synthesis — MABAL

MABAL (Module and Bus Allocation) simultaneously allocates and binds functional units, registers, and interconnect, trying to minimize total cost by trading-off between module cost and interconnect cost.

Examples

MAHA

An example from Park, and a temperature controller example.

REAL

Two dataflow examples.

MABAL

A controller example, a fifth-order digital elliptic wave filter example, and a two-stage multiplier

References

Küçükçakar90a

Kayhan Küçükçakar and Alice C. Parker, "Data Path Tradeoffs using MABAL", *Proc. of the 27th DAC*, pages 511–516, June 1990.

Data path synthesis, and a controller example.

Küçükçakar90b

Kayhan Küçükçakar and Alice C. Parker, "MABAL: A Software Package for Module and Bus Allocation", *International Journal of Computer Aided VLSI Design*, pages 419–436, 1990.

ADAM overview, data path synthesis, a fifth-order digital elliptic wave filter example, and a two-stage multiplier.

Afsarmanesh89

Hamideh Afsarmanesh, Esther Brotoatmodjo, Kwang June Byeon, and Alice C. Parker, "The EVE VLSI Information Management Environment", *Proc. of ICCAD'89*, pages 384–387, November 1989.

An object-oriented database framework.

Jain89

Rajiv Jain, Kayhan Küçükçakar, Mitchell J. Mlinar, and Alice C. Parker, "Experience With the ADAM Synthesis System", *Proc. of the 26th DAC*, pages 56–61, June 1989.

Updated system overview, including prediction and synthesis, and an AR lattice filter example.

Hayati89

Sally Hayati and Alice Parker, "Automatic Production of Controller Specifications From Control and Timing Behavioral Descriptions", *Proc. of the 26th DAC*, pages 75–80, June 1989.

Generating controller specifications from high-level control and timing graphs, and a multi-bus slave example.

Kurdahi89

Fadi J. Kurdahi and Alice C. Parker, "Techniques for Area Estimation of VLSI Layouts", *IEEE Trans. on CAD*, pages 81–92, January 1989.

Area estimation for standard cell layout, including logic area, wiring area, and pad area.

Jain88a

Rajiv Jain, Mitchell J. Mlinar, and Alice C. Parker, "Area-Time Model for Synthesis of Non-Pipelined Designs", *Proc. of ICCAD'88*, pages 48–51, November 1988.

Predicting the lower-bound area-time tradeoff curve, considering functional units, but not considering registers or interconnection.

Hayati88

Sally A. Hayati, Alice C. Parker, and John J. Granacki, "Representation of Control and Timing Behavior with Applications to Interface Synthesis", *Proc. of ICCD'88*, pages 382–387, October 1988.

The Design Data Structure (DDS).

Jain88b

Rajiv Jain, Alice Parker, and Nohbyung Park, "Module Selection for Pipelined Synthesis", *Proc. of the 25th DAC*, pages 542–547, June 1988.

Generating, ranking, and choosing module sets for pipelined designs.

Park88

Nohbyung Park and Alice C. Parker, "Sehwa: A Software Package for Synthesis of Pipelines from Behavioral Specifications", *IEEE Trans. on CAD*, pages 356–370, March 1988.

Sehwa overview, pipeline synthesis, resource allocation, and scheduling.

Jain87

Rajiv Jain, Alice Parker, and Nohbyung Park, "Predicting Area-Time Tradeoffs for Pipelined Design", *Proc. of the 24th DAC*, pages 35–41, June 1987.

Predicting cost/speed tradeoffs for pipelined designs, including operator, register, and multiplexor estimation.

Kurdahi87

Fadi J. Kurdahi and Alice C. Parker, "REAL: A Program for REgister ALlocation", *Proc. of the 24th DAC*, pages 210–215, June 1987.

Register binding, an example from Park, and a temperature controller example.

Granacki87

John J. Granacki, Jr. and Alice C. Parker, "PHRAN-SPAN: A Natural Language Interface for System Specifications", *Proc. of the 24th DAC*, pages 416–422, June 1987.

A natural language interface to the Design Data Structure..

Knapp86a

David W. Knapp, *A Planning Model of the Design Process*, PhD Thesis, EE Dept., Univ. of Southern California, December 1986.

The Design Data Structure (DDS) and planning.

Granacki86

John Joseph Granacki, Jr., *Understanding Digital System Specifications Written in Natural Language*, PhD Thesis, EE Dept., Univ. of Southern California, December 1986.

Using a natural language input to specify asynchronous concurrent behavior (PHRAN-SPAN).

Afsarmanesh86

Hamideh Afsarmanesh, Dennis McLeod, David Knapp, and Alice C. Parker, "Information Management for VLSI/CAD", *Proc. of ICCD'86*, pages 476–481, October 1986.

Overview of the 3 Dimensional Information Space (3DIS) information management framework.

Knapp86b

David W. Knapp and Alice C. Parker, "A Design Utility Manager: the ADAM Planner", *Proc. of the 23rd DAC*, pages 48–54, June 1986.

Building a design plan, estimation, the design knowledge base.

Park86

Nohbyung Park and Alice C. Parker, "Sehwa: A Program for Synthesis of Pipelines", *Proc. of the 23rd DAC*, pages 454–460, June 1986.

Sehwa overview, pipeline synthesis, resource allocation, and scheduling.

Parker86

Alice C. Parker, Jorge "T" Pizarro, and Mitch Mlinar, "MAHA: A Program for Datapath Synthesis", *Proc. of the 23rd DAC*, pages 461–466, June 1986.

Scheduling, functional unit binding, and two dataflow examples.

Kurdahi86

Fadi J. Kurdahi and Alice C. Parker, "PLEST: A Program for Area Estimation of VLSI Integrated Circuits", *Proc. of the 23rd DAC*, pages 467–473, June 1986.

Estimating standard cell layouts using a probabilities model for cell placement and interconnection.

Park85

Nohbyung Park and Alice C. Parker, "Synthesis of Optimal Clocking Schemes", *Proc. of the 22nd DAC*, pages 489–495, June 1985.

Choosing the minimum and optimal clock period to maximize the execution speed, choosing the optimal number of clock phases per cycle, inserting storage elements, and assigning operations to clock phases.

Granacki85

John Granacki, David Knapp, and Alice C. Parker, "The ADAM Advanced Design Automation System: Overview, Planner and Natural Language Interface", *Proc. of the 22nd DAC*, pages 727–730, June 1985.

System overview, the Design Data Structure, the design planner, and the natural language interface.

Parker84

Alice C. Parker, Fadi Kurdahi, and Mitch Mlinar, "A General Methodology for Synthesis and Verification of Register-

Transfer Designs", *Proc. of the 21st DAC*, pages 329–335, June 1984.

Synthesis techniques, and the Design Data Structure.

Univ. of Texas at Austin's DAGAR System

The DAGAR system was Raj's thesis work at the University of Illinois at Urbana-Champaign, and is being continued at the University of Texas at Austin. DAGAR includes scheduling and data path synthesis.

Input

Hand-generated dataflow graph.

Internal Behavioral Representation

A dataflow graph.

Scheduling / Data Path Synthesis

First, an ASAP schedule is constructed, assuming infinite resources, and one cycle per operation. Then optimizations are applied, moving operations to other control steps to reduce the maximum number of operations of each type in any one control step, and grouping operations into functional units so as to have a minimum number of functional units. Then the scheduler traverses the control step schedule, passing the operations in each control step to the data path allocator. The data path allocator tries to bind those operations using heuristics; if it fails, the scheduler tries to delay operations until later control steps, and if that also fails, the user is notified that the resource constraints should be increased.

Examples

An integer restoring division algorithm example.

References

Raj89

Vijay K. Raj, "DAGAR: An Automatic Pipelined Microarchitecture Synthesis System", *Proc. of ICCD'89*, pages 428–431, October 1989.

Scheduling and data path synthesis, and an integer restoring division algorithm example.

Raj86

Vijay K. Raj "Another Automated Data Path Designer", *Proc. of ICCAD'86*, pages 214–217, November 1986.

Scheduling and data path synthesis.

Univ. of Waterloo's CATREE System

The University of Waterloo's CATREE (Computer aided tree) system includes scheduling and data path synthesis.

Input

An algorithmic behavioral specification, describing only straight-line code segments.

Internal Behavioral Representation

A directed acyclic operator graph.

Scheduling

ASAP and ALAP scheduling.

Data Path Synthesis

Builds a binary tree, placing operators into the tree according to their degree of shared variable connectivity, with more connected operators closer to the root of the tree. Bottom-up tree traversal algorithms are used to cluster operators and values and perform functional unit allocation and register allocation, respectively. A top-down tree traversal algorithm is used to assign interconnect.

Examples

A fifth-order digital elliptic wave filter example.

References

Gebotys88

Catherine H. Gebotys and Mohamed I. Elmasry, "VLSI Design Synthesis with Testability", *Proc. of the 25th DAC*, pages 16–21, June 1988.

System overview, scheduling, data path synthesis, and a fifth-order digital elliptic wave filter example.

Univ. of Waterloo's VLSI CAD Tools

The University of Waterloo's VLSI CAD Tools include ACE, a graphical interface, and SPAID, a design-specific system for DSP synthesis, including transformations, scheduling, and data path synthesis.

Input

Uses the ACE graphical interface.

Internal Behavioral Representation

A Generalized Signal Flow Graph (GSFG), where each node represents an operation and its corresponding delay, and each edge represents data dependencies between operations.

Transformations

Systolic transformations, plus local transformations (associative and distributive transformations, elimination of redundant operations, etc.).

Scheduling

Phase 1

Groups the operators into classes of data independent operators, using a linear programming technique.

The user can alter this scheduling by weighting individual operations, or by adding extra precedence edges to the GSFG.

Phase 2

Orders operations within each class, using a greedy heuristic. Operations that take longer to execute are scheduled before operations that take less time, and for operations that take an equal amount of time, the operation with the earliest available data is scheduled.

Data Path Synthesis

Functional Unit and Bus Binding

Uses a greedy heuristic, choosing the functional unit that will result in the least idle time. At the end of this phase, all buses are connected to all inputs of all functional units via multiplexors, and all registers are connected to all buses.

Register Grouping and Multiplexor Minimization

Groups registers into one register file per bus, using a bipartite graph coloring to insure the minimum number of buses. Heuristics are also used to minimize the number of multiplexor inputs.

Register Minimization

Uses a graph coloring algorithm to minimize the number of registers in each register file.

Examples

A 16-point FIR filter example, a fifth-order digital elliptic wave filter example, and a fifth-order WDF/CIR filter example.

References

Gebotys90

Catherine H. Gebotys and Mohamed I. Elmasry, "A Global Optimization Approach for Architectural Synthesis", *Proc. of ICCAD'90*, pages 258–261, November 1990.

A linear programming approach to scheduling and allocation, a fifth-order digital elliptic wave filter example, and a differential equation example.

Mahmood90

M. Mahmood, F. Mavaddat, and M.I. Elmasry, "A Formal Approach to Control-Unit Synthesis", *Proc. of the IFIP Working*

Conf. on Logic and Architecture Synthesis, pages 126–135, May, 1990.

Describing the data path and control unit with formal Register-Transfer level models, mapping the formal specification onto microcode, and generating a state-table representation of the control unit.

Mahmood89

M. Mahmood, F. Mavaddat, M.I. Elmasry, and M.H.M. Cheng, "A Formal Language Model of Local Microcode Synthesis", *Proc. of the IMEC-IFIP Workshop on Applied Formal Methods for Correct VLSI Design*, pages 21–39, November, 1989.

Describing the data path with a formal Register-Transfer level model, mapping the formal specification onto microcode, a second-order differential equation example, and Tseng's running example.

Buset89

O.A. Buset and M.I. Elmasry, "ACE: A Hierarchical Graphical Interface for Architectural Synthesis", *Proc. of the 26th DAC*, pages 537–542, June 1989.

The ACE graphical interface, and a third-order digital elliptic wave filter example.

Haroun89

Baher S. Haroun and Mohamed I. Elmasry, "Architectural Synthesis for DSP Silicon Compilers", *IEEE Trans. on CAD*, pages 431–447, April 1989.

Design style selection, transformations, self-timed data communication, scheduling, data path synthesis, a 16-point FIR filter example, a fifth-order digital elliptic wave filter example.

Haroun88

B.S. Haroun and M.I. Elmasry, "Automatic Synthesis of a Multi-Bus Architecture for DSP", *Proc. of ICCAD'88*, pages 44–47, October 1988.

Scheduling, data path synthesis, a fifth-order digital elliptic wave filter example, and a fifth-order WDF/CIR filter example.

Index